Materials Research Needs
to Advance Nuclear Energy

T0306002

MATERIALS RESEARCH SOCIETY
SYMPOSIUM PROCEEDINGS VOLUME 1215

Materials Research Needs to Advance Nuclear Energy

Symposium held November 30–December 4, Boston, Massachusetts, U.S.A.

EDITORS:

Gianguido Baldinozzi

Materiaux Fonctionnels pour l'Energie
Chatenay-Malabry, France

Yanwen Zhang

Pacific Northwest National Laboratory
Richland, Washington, U.S.A.

Katherine L. Smith

Nuclear Science and Technology Embassy of Australia,
Washington, D.C., U.S.A.

Kazuhiro Yasuda

Kyushu University
Fukuoka, Japan

Materials Research Society
Warrendale, Pennsylvania

CAMBRIDGE
UNIVERSITY PRESS

University Printing House, Cambridge CB2 8BS, United Kingdom

One Liberty Plaza, 20th Floor, New York, NY 10006, USA

477 Williamstown Road, Port Melbourne, VIC 3207, Australia

314-321, 3rd Floor, Plot 3, Splendor Forum, Jasola District Centre, New Delhi - 110025, India

79 Anson Road, #06-04/06, Singapore 079906

Cambridge University Press is part of the University of Cambridge.

It furthers the University's mission by disseminating knowledge in the pursuit of education, learning and research at the highest international levels of excellence.

www.cambridge.org
Information on this title: www.cambridge.org/9781605111889

Materials Research Society
506 Keystone Drive, Warrendale, PA 15086
http://www.mrs.org

© Materials Research Society 2010

First published 2010
First paperback edition 2012

Single article reprints from this publication are available through University Microfilms Inc., 300 North Zeeb Road, Ann Arbor, MI 48106

CODEN: MRSPDH

A catalogue record for this publication is available from the British Library

ISBN 978-1-605-11188-9 Hardback
ISBN 978-1-107-40806-7 Paperback

CONTENTS

* Invited paper.

* Invited paper.

Modeling Complex Materials II

Poster Session: Advanced Materials for Nuclear Energy

Metallic Materials II

Complex Materials and Devices

PREFACE

This volume contains papers presented in Symposium V, "Materials Research Needs to Advance Nuclear Energy," held on November 30–December 4 at the 2009 MRS Fall Meeting in Boston, Massachusetts. The purpose of this symposium was to bring together experimenters, theoreticians and modelers to discuss the innovations needed to develop the next generation of nuclear materials, and to understand the performance of existing materials under extreme operating conditions. The symposium included cross-cutting sessions on radiation effects, complex microstructures, metallic materials, ceramic materials and waste forms, modeling of complex materials, chemical and structural complexity of advanced fuels, and complex materials and devices. Over 130 vibrant cutting-edge presentations were given by researchers from universities, national laboratories and industries from 14 countries for a period of four and half days, including 19 invited talks and 71 contributed talks together with over 40 posters. This volume contains 27 papers presented in this symposium. These presentations explored the fabrication (melting, rolling sol gel, sintering, hot-pressing), characterization (microscopy, diffraction, thermal and electrical property measurements), modeling (ranging from nano- to mesoscale, and spanning timeframes ranging from fractions of femtoseconds to hundreds or millions of years), and performance prediction of various nuclear fuel cycle materials.

The organizers and editors of these symposium proceedings would like to thank all the people who have contributed towards making this symposium happen, the session chairs, the symposium technical and editorial staff and, above all, the participants and the authors for helping to make this symposium successful and fruitful.

Gianguido Baldinozzi
Yanwen Zhang
Katherine Smith
Kazuhiro Yasuda

June 2010

MATERIALS RESEARCH SOCIETY SYMPOSIUM PROCEEDINGS

MATERIALS RESEARCH SOCIETY SYMPOSIUM PROCEEDINGS

Prior Materials Research Society Symposium Proceedings available by contacting Materials Research Society

Complex Microstructures

Mater. Res. Soc. Symp. Proc. Vol. 1215 © 2010 Materials Research Society   1215-V02-01

Can We Describe Phase Transition in Insulators within the Landau PT Theory Framework?

David Simeone[1], Gianguido Baldinozzi[2], Dominique Gosset[1], Laurence Luneville[3], and Leo Mazerolles[4]

[1]CEA/DEN/DMN/SRMA/LA2M, Matériaux Fonctionnels pour l'Energie, Equipe Mixte CEA-CNRS-ECP, CEN Saclay, Gif sur Yvette, 91191, France.
[2]CNRS, Matériaux Fonctionnels pour l'Energie, CNRS-CEA-ECP, Laboratoire SPMS, Ecole Centrale de Paris, Chatenay Malabry, 92292, France.
[3]CEA/DEN/DM2S/SERMA/LLPR, Matériaux Fonctionnels pour l'Energie, Equipe Mixte CEA-CNRS-ECP, CEN Saclay, Gif sur Yvette, 91191, France.
[4]CNRS, Institut des Sciences Chimiques Seine Amont, Thiais, 92000, France.

ABSTRACT

Based on studies of simple oxides, this paper demonstrates that the specific energy deposition modes under irradiation induce modifications of materials over different length scales. On the other hand, we show the Landau phase transition theory, widely used to explain the structural stability of materials out of irradiation, can give a general framework to describe the behavior of these oxides under irradiation. The use of X-ray diffraction techniques coupled with the Raman spectroscopy allows defining in a quantitative way the phenomenological parameters leading to predictive results. This paper clearly shows that in two model systems, pure zirconia and spinels, no unexpected new phases are produced in these oxides irradiated at room temperature and with different fluxes. Such a phenomenological approach may be useful to study the radiation tolerance of many crystalline ceramics (e.g. the zirconium based americium ceramics).

INTRODUCTION

A significant growth in the utilization of nuclear power is expected to occur over the next several decades due to increasing demand for energy and environmental concerns related to emissions from fossil fuel plants. This has focused increasing attention on issues related to the permanent disposal of nuclear waste and the improvement of nuclear plants technologies. The achievement of such a goal requires continued improvements in safety and efficiency particularly related to the performance of materials [1]. The related research challenges represent some of the most demanding tests of our fundamental understanding of materials science and chemistry, and they provide significant opportunities for advancing basic science with broad impacts in the material science community. The fundamental challenge is to understand and control chemical and physical phenomena in multi-component systems from femto-seconds to millennia, at temperatures up to 1000°C, and for radiation doses up to hundreds of displacements per atom (dpa). New understanding is required for the microstructural evolution of the materials. Their phase stability under irradiation is an active field of debate.

In this context, ceramics appear to be incontrovertible materials. Crystalline oxide waste forms such as zircon [2] or spinel [3] have been proposed to accommodate a limited range of active species such as plutonium and multiphase systems such as Synroc [4] to accommodate a broader range of active species. On the other hand, carbides are promising candidates for claddings or diffusion barriers in the next generation of nuclear plants. The predictive understanding of microstructural evolutions and property changes of these ceramics under high temperature and elevated dpa is essential for the rational design of materials for structural, fuels, and waste-form applications. A clear relationship between energy depositions by impinging particles, phase stability and mechanical behavior has to be achieved.

The Phase Transition Landau framework remains a powerful tool to describe the evolution of solids submitted to various conditions (temperature, pressure) [6]. Variation of order parameters, concentration waves and strain field can be analyzed with the same formalism in term of minimization of an "effective Hamiltonian" within the mean field approximation [6,7,8]. Numerous authors try to analyze the behavior of alloys under irradiation within this framework [9,10]. A question naturally arises: can we apply this formalism to describe and to capture the main features of radiation induced phase transition in ceramics and in a more general way structural evolution of ceramics under irradiation?. The answer to such a question is not simple. In condensed matter, the control parameter (temperature or pressure) in the Landau Free energy is uniform along the material. In radiation experiments, the control parameter associated with the sub cascades formation describing the energy deposition is not uniform. On the other hand, fluctuations of order parameters induced by irradiation can be expected to be violent. A "Landau effective Hamiltonian" could thus not be able to describe the physics of radiation damage at least in alloys [11,12]. Such a harsh treatment leads to the appearance of new unexpected phases under irradiation in alloys [12].

Based on two examples of phase transition in simple oxides irradiated by different ions, we discuss in this paper the validity of such an approach. Two different oxide, pure zirconia and magnesium spinels were irradiated at room temperature with different ions. Their structural evolutions were explained within the Landau theory of phase transition pointing out that such formalism clearly describes the patterning observed in these irradiated ceramics as a function of few well defined experimentally measurable parameters.

EXPERIMENT

The main difficulty to understand the long term behavior of materials under irradiation is due to the fact that radiations damages occur over different length scales. The characteristic penetration depth of Primary Knock on Atoms or incident ions is of about few hundreds of nanometers to tens of micrometers, a highly disordered area is created over few nanometers in thermal spikes and point defects are produced at the atomic scale. Many experimental tools, like optical spectroscopy, positron annihilation and electron paramagnetic resonance have been developed to track point defects and aggregates induced by irradiation in insulators and semi conductors [13]. Few techniques are able to probe the nanometric scale. Rutherford back scattering channeling [14] quantifies the modification of atomic rows due to radiation damage at the nanometric scale but cannot give a clear description of e.g. modifications of the space group associated with the crystal lattice. Despite Electron Microscopy remains a powerful tool to track structural modifications of solids induced by irradiation, an accurate analysis of the diffraction patterns remains difficult. X ray gives valuable information at different scales. The kinematic approximation permits to describe the atomic structure of the material after irradiation: the Bragg peaks modifications allow describing the evolution of the space group associated with the atomic lattices and the effect of strains induced by irradiation can be measured from the analysis of the lines broadening. On the other hand, Raman spectroscopy gives valuable information on the local environment. These two techniques provide measureable ingredients to describe radiation damages within the Landau framework as pointed out by many authors working on ceramics out and under irradiation [6,8,15,16]

We now illustrate this point describing the evolution of two simple oxides under irradiation.

Behavior of pure zirconia irradiated by low and high energy ions

Many authors[17,18] have shown that pure zirconia exhibits a first order phase transition as a function of the temperature. As irradiation is able to strongly modify the crystal, we can then enhance or inhibit this phase transition under irradiation. To assess this point, various irradiations were performed on pure monoclinic zirconia samples at room temperature [15]. Figure 1 exhibits the Raman spectra (left figure) collected on pure zirconia sample after irradiation by 800 keV Bi ions with a fluence of 10^{15} cm^{-2} at room temperature. The existence of a large Raman peak at about 250 cm^{-1} demonstrates that the tetragonal phase is produced under irradiation in pure monoclinic zirconia at room temperature. From this experimental result, X ray diffraction technique was used both to quantify the amount of the tetragonal phase formed under irradiation (right figure) and the strain field associated with this new phase [15].

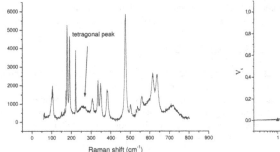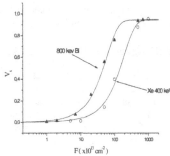

Figure 1. Raman Evidence of the existence of a tetragonal phase induced by 800 KeV Bi impinging ions at room temperature in pure zirconia (left). The appearance of Raman peak at around 260 cm^{-1} (arrow) and the absence of a Raman peak near 600 cm-1 assess that only the tetragonal phase of zirconia is produced. The Right figure displays and quantitative measure of the amount of the tetragonal phase , noticed Vt, versus the fluence F, for different irradiations performed at room temperature. The amount of the tetragonal phase results from a Rietveld refinement of the diffraction patterns collected

under grazing incidence [15].

Figure 2 exhibits the evolution of the strain field as a function of the annealing temperature on the most irradiated sample of pure zirconia irradiated by 92 MeV Kr ions at room temperature [19]. The analysis of Bragg peaks gives quantitative information on the strain tensor.

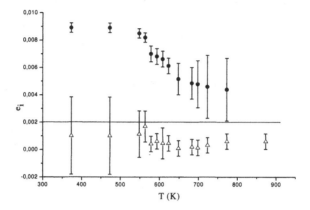

Figure 2. Evolution of the different components of the strain tensor (I varies from 1 to 3 in Voigt notation) extracted from the broadening of Bragg peaks versus the temperature during isochronal annealings on zirconia samples irradiated by 92 MeV Kr ions [18].

Combined analysis of Raman and XRD measurement will then permit to draw a clear picture describing the evolution of zirconia under irradiation [19,20].

Behavior of spinels irradiated by low and high energy ions

The impact of radiation has also been studied on magnesium aluminates, magnesium chromites and zinc aluminates spinels. This family exhibits a very simple phase diagram out of irradiation. These samples were irradiated at room temperature by both high and low energy ions to detect a possible phase transition induced by irradiation. The damaged are were then scanned using both Raman spectroscopy and XRD. Figure 3 displays the evolution of XRD patterns and Raman spectra on $MgCr_2O_4$ samples irradiated by 92 MeV Kr ions [18].

The XRD diffraction patterns clearly exhibit that many Bragg peaks vanish on the most irradiated samples whereas Raman analysis shows all normal modes associated with the usual spinel space group (Fd-3m) remain. The vanishing of these odd Bragg peaks could be understood as a phase transformation induced by irradiation at the atomic scale. However, TEM patterns collected on different scales give the experimental evidence that no change of the space group occurs in irradiated spinels in agreement with results of Raman spectroscopy which does not reveal any structural change at this scale. Figure 3 highlights the fact that irradiation can modify materials over different scale lengths. Moreover, no clear broadening is observed on odd Bragg peaks insuring that no strain appears during irradiation in the materials.

6

Since intensities of Raman peaks do not evolve during irradiation, the stiffness of bonds remains unaffected by irradiation. At first approximation, Raman frequencies can be considered as a simple function of the average mass of the cation lying on the octahedral and tetrahedral sites. Assuming the stiffness constant of the cation oxygen bond unaffected by irradiation, it is then possible to estimate the fraction of Mg atoms lying at the center of octahedral sites. From the measure of the Raman shifts, this fraction of Mg atoms in octahedral positions is roughly equal to 20% in the most irradiated samples. This point has been discussed in detail [23].

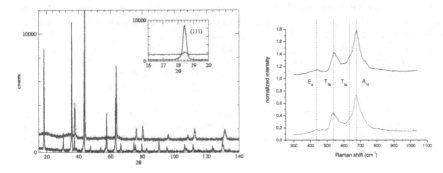

Figure 3. Comparison between Raman spectra (right figure) and XRD patterns collected on non irradiated (black line) and on the most irradiated (red line) sample of $MgCr_2O_4$.

DISCUSSION

All information given by Raman spectroscopy and X ray diffraction collected on irradiated samples can be clearly described with the phase transition Landau theory framework. Whereas this theory remains phenomenological, it can be used to describe the microstructural evolution of ceramics under irradiation taken account all experimental facts. We illustrate this point below.

Structural evolution of pure zirconia under irradiation

The phase transition Landau theory gives the natural framework to describe evolution of these ceramics as a function of the temperature and the pressure out of irradiation. Some authors have clearly shown that the monoclinic to tetragonal phase transition occurring in pure zirconia can be described by a phonon condensation [22]. The description of this transition within the Landau theory agrees with experimental results. Moreover, this description explains the stability of the tetragonal phase, the high temperature phase, in nanocrystals of pure zirconia at room temperature [23], pointing out the role of the strain field induced by the surface free energy. Such a model explains also the appearance and the stability of the tetragonal phase under irradiation [24]. Neglecting Ginzburg terms in Landau free energy and focusing our attention only on the leading terms, it is possible to write the free energy density as:

$$F(I) = \frac{\alpha_T}{2}\left(T - T_c - \frac{\alpha_e}{\alpha_T}H\left(e - e_{Threshold}\right)\right)I - \frac{1}{4}I^2$$

where α_T and α_e are phenomenological positive constants. The function I is the Landau invariant responsible for the displacive transition out of irradiation and 'e' is the trace of the strain tensor.

The analysis of the stability of zirconia nanocrystals as a function of the temperature out of irradiation and XRD patterns of irradiated zirconia samples give experimental evidence to include this term in the Landau free energy [23,24]. The value α_e embodies then the correction to the free energy density associated with the strain field induced by oxygen vacancies (deduced from optical analysis) in low energy irradiation and dislocation loops (as observed by TEM) in high energy irradiations. Once the concentration of defects is high enough, the strain fields becomes superior to a threshold value and the tetragonal phase is stabilized. This point clearly shows that radiation damages can be easily integrated in the PT Landau formalism.

Moreover, it is important to notice that Bright field TEM pictures clearly display the fact that the tetragonal domains produced under irradiation are not uniformly distributed in the irradiated sample. This point highlights the fact that different length scales exist under irradiation as expected from the fractal feature of the energy deposition under irradiation. [16,17]..

Structural evolution of spinels under irradiation

Out of irradiation, many authors attempted to describe the cation exchange occurring as a function of the temperature in many spinels within the mean field approximation. Careful analyses of neutron diffraction data prove that this thermodynamic of non convergent cation ordering in spinels can be better described within the Landau phase transition framework [7]:

$$F(I) = -hI + \frac{\alpha_T}{2}\left(T - T_c\right)I^2 + \frac{c}{6}I^6$$

Where I is the order parameter related to the cation exchange along tetrahedral and octahedral sites. The constants h, α_T and c are positive phenomenological constants.

This Landau free energy explains the evolution of the cation exchange as a function of the temperature out of irradiation without evoking any strain field effects, this in agreement with XRD results (no Bragg peaks broadening). Such a free energy can also explain the behavior of these materials maintained far from equilibrium by irradiation. The analysis of Raman spectra on spinels irradiated by low and high energy ions clearly displays the appearance of different shoulders in irradiated samples indicating andifferent domains characterized by different values of the order parameter I. This point can be understood if we correlate these domains with existence of displacement cascades induced by irradiation in spinels as pointed out by Molecular Dynamic simulation [24].

It is important to notice that XRD patterns clearly reveal the fact that the domains produced under irradiation are not uniformly distributed in the irradiated sample. As pointed out in our previous example, this point highlights the fact that different length scales exist under irradiation in agreement with the fractal nature of the energy deposition [14,15].

CONCLUSIONS

As pointed out by numerous authors [25-27], the state of the knowledge on the behavior of oxides and more generally of ceramics under irradiation is far from satisfactory, despite their important role in these conditions, either as electric insulators, as nuclear fuels and structural materials in reactors or as storage matrices for radioactive wastes from nuclear industry. The origin of this situation can be attributed to:

- Experiments in ceramics are more complex than in metals and alloys due to the higher characteristic temperatures involved.
- the complex nature of the bonding, partly ionic [24] and partly covalent [27],in ceramics are less known than in alloys. This point has for long precluded any relevant quantitative calculation.

Based on studies of two simple oxides, pure zirconia and spinels, this paper demonstrates that the specific energy deposition modes induce modifications of materials over different length scale. On the other hand, we prove that the Landau phase transition theory largely used to explain the structural stability of materials out of irradiation can give a general framework to describe the behavior of these oxides under irradiation. The major point of this paper is clearly to show that no new phases, unexpected from the classical P-T thermodynamic diagram of these solids, are produced under irradiation with different fluxes at least at room and low temperatures.

Moreover, the Phase Transition Landau theory succeeded to explain the behavior of these materials under irradiation. Even if this theory remains phenomenological, the use of diffraction techniques coupled with the Raman spectroscopy allows defining in a quantitative way parameters leading to predictive results. On the other hand, this framework can be easily extended to take into account inhomogeneities due to irradiation in larger length and time scales. Even if this description of radiation damage remains phenomenological, it gives some clues to capture key parameters controlling the microstructure of these irradiated simple oxides, overcoming difficulties associated with the complex nature of chemical bonds in ceramics. Such a phenomenological approach may be used to understand the radiation resistance of many crystalline ceramics under irradiation like for instance the zirconium based americium ceramics [30] or the amorphisation of SiC [5].

REFERENCES
[1] Jim Roberto, Thomas diaz de la Rubbia, "Basic research needs for advanced nuclear energy systems, DOE report (2006).
[2] K. E. Sickafus, L. Minervini, R. W. Grimes, et al.: Science **289,** 748 (2000).
[3] K. E. Sickafus, A. C. Larson, N. Yu, et al.: J.M.N. **219,** 128 (1995).

[4] Ringwood, A.E., Kesson, S.E., Ware, N.G., Hibberson, W.O. and Major, A. Immobilization of high level nuclear reactor wastes in SYNROC. *Nature* **278**, 219-223 (1979).
[5] W. Weber, L. Wang, Y. Zhang, W. jiang, I. Bae, NIMB 266, 2797(2008)
[6] P. Toledano, V. Dimitriev, Reconstructive Phase Transition, World Scientific, 1996.
[7] M Carpenter, E. Salje, Amer. Min. 79, 1068 (1994).
[8] E Salje, Phys. Chem. Min. 15, 336 (1988).
[9] for a review on these problems see F Ducastelle, order and phase stability in alloys, NH 1991.
[10] for a review on these problems see A. Katchaturyan, The theory of structural Transformations in Solids , Wiley, 1983.
[11] G. Martin, P. Bellon, Sol. State. Phys., 53--54, 1 (1997).
[12] for a recent review on these problems see G. Martin, P. Bellon, CR physique 9, 323 (2008).
[13] A Dunlop, RF Rullier-Albenque, C. jaouen, C templier, J. davenas, Materials under irradiation, trans Tech Publications 1993.
[14] D. Simeone, L. Luneville and J. P. Both, EPL, 83 (2008) 56002
[15] D. Simeone, L. Luneville, Phys. Rev. E **81**, 021115 (2010)
[16] L. Feldman; J. Mayer, S. Picraux). *Materials Analysis by Ion Channeling*, Academic Press 1982.
[17] D. Simeone, J. L. Bechade, D. Gosset, et al.: J. Nuc. Mater. **281**, 171 (2000)
[18] D. Simeone, C. Dodane-Thiriet, D. Gossetl.: J. Nuc. Mater. **300** 151 (2002).
[19] R. Patil and E. Subbarao, Acta Crystallogr., Sect. A: Cryst. Phys., Diffr., Theor. Gen. Crystallogr. **26**, 535 (1970).
[20] D. Simeone, G. Baldinozzi, D. Gosset.: Phys. Rev. B 70 (2004), 134116.
[21] G. Baldinozzi, D Simeone, D Gosset, I. Monnet,S. Le Caër and Léo Mazerolles, Phys. Rev. B 74, 132107 _(2006).
[22] D. Simeone, G. Baldinozzi, D. Gosset, S. LeCaër, and L. Mazerolles, Phys. Rev. B **70**, 134116 (2004)
[23] D. Simeone, G. Baldinozzi, D. Gosset, L. Mazerolles, and L. Thome Mater. Res. Soc. Symp. Proc. Vol. 1043 (2008).
[24] A. Chartier, C. Meis, J.P. Crocombette, W. Weber, L. Corrales, Phys. Rev. Lett. 94, 25505 (2005).
[25] G. P. Pells: J.M.N. **155-157**, 67 (1988)
[26] G. P. Pells and D. C. Phillips: J. Mat. Nuc. **80**, 207 (1979).
[27] G. P. Pells: Rad. Eff. **64**, 71 (1982).
[28] F. Clinard and L. Hobbs: in Modern problems in condensed matter physics, North-Holland, Amsterdam (1986).
[29] J. C. Bourgoin and J. W. Corbett: Rad. Eff. **36**, 157 (1978).
[30] R Belin, P Martin, P Valenza, A Scheinos, Inorg chem.. 48, 5376 (2009).

Mater. Res. Soc. Symp. Proc. Vol. 1215 © 2010 Materials Research Society   1215-V02-04

HRTEM Study of Oxide Nanoparticles in 16Cr-4Al-2W-0.3Ti-0.3Y$_2$O$_3$ ODS Steel

Luke L. Hsiung[1], Michael J. Fluss[1], Mark A. Wall[1], Akihiko Kimura[2]

[1]Lawrence Livermore National Laboratory
Physical and Life Sciences Directorate
L-352, P.O. Box 808
Livermore, CA, U.S.A.

[2]Institute of Advanced Energy
Kyoto University, Gokasho, Uji
Kyoto 611-0011, Japan

ABSTRACT

Crystal and interfacial structures of oxide nanoparticles in 16Cr-4Al-2W-0.3Ti-0.3Y$_2$O$_3$ ODS steel have been examined using high-resolution transmission electron microscopy (HRTEM) techniques. Oxide nanoparticles with a complex-oxide core and an amorphous shell were frequently observed. The crystal structure of complex-oxide core is identified to be mainly Y$_4$Al$_2$O$_9$ (YAM) with a monoclinic structure. Orientation relationships between the oxide and matrix are found to be dependent on the particle size. Large particles (> 20 nm) tend to be incoherent and have a spherical shape, whereas small particles (< 10 nm) tend to be coherent or semi-coherent and have faceted interfaces. The observations of partially amorphous nanoparticles lead us to propose a three-stage mechanism to rationalize the formation of oxide nanoparticles containing core/shell structures in as-fabricated ODS steels.

INTRODUCTION

Development of high-performance structural materials for first wall and breeding-blanket components, which will be exposed to high fluxes of high energy (14 MeV) neutrons from the deuterium-tritium fusion, is one of the major challenges in materializing future fusion reactors. The choice of structural materials for the first wall and blanket to a large degree dictates the design of the reactor systems. The selection of suitable structural materials is based on conventional properties (such as thermophysical, mechanical, and corrosion and compatibility), low neutron-induced radioactivity, and resistance to radiation-induced damage phenomena like material hardening/embrittlement and/or dimensional instability caused by void- and helium-driven swelling [1]. Oxide dispersion strengthened (ODS) F/M and ferritic steels, which produced by mechanical alloying the elemental (or pre-alloyed) metallic powder and yttria (Y$_2$O$_3$) oxide powder consolidated by hot extrusion or hot isostatic pressing, are advanced structural materials with a potential to be used at elevated temperatures due to the dispersion of thermally stable oxide nanoparticles into the F/M matrix. The use of ODS steels should improve creep strength and oxidation/corrosion resistance at high temperatures and consequently increase the operating temperature of first wall and blanket structures in the future fusion/fission hybrid reactors by approximately 700 °C or higher [2]. The performance of ODS steels is largely determined by the particle size and the stability of dispersed oxide nanoparticles. Although Y$_2$O$_3$ has been selected as the major dispersed oxide, its particle size increases during the consolidation and thermomechanical treatment of ODS steels. To enhance the stability of oxide particles, titanium and aluminum are added to form complex oxide with Y$_2$O$_3$ so as to make the dispersed oxides finer and more stable [2, 3]. Complex mechanisms involving fragmentation, dissolution, and precipitation of complex-oxide nanoparticles were previously proposed by Okuda and Fujiwara [4], Kimura et al. [5] and Sakasegawa et al. [6] based on the results generated from x-ray diffraction and conventional TEM studies. However, recent studies conducted by Marquis [7] using 3-D atom probe tomography and Klimiankou et al. [8] using EDX and EELS methods have revealed the existence of a complex-oxide core structure associated with a solute-enriched shell structure in oxide nanoparticles. The existence of core/shell structures in nanoparticles however challenges the accountability of previously proposed dissolution/precipitation mechanisms. HRTEM study of nanoparticles in K3-ODS was thus conducted to examine the crystal and interfacial structures of oxide nanoparticles to better understand the formation mechanisms of oxide nanoparticles associated with core/shell structures.

EXPERIMENTAL

The material used for this investigation was 16Cr-4Al-2W-0.3Ti-0.3Y$_2$O$_3$ (K3) ODS ferritic steel [9]. Details of the fabrication process of the ODS steels can be found elsewhere [10]. Briefly, the pre-alloyed powder was first mechanically alloyed with Y$_2$O$_3$ powder in an Argon gas atmosphere at room temperature using an attrition type ball mill. The powder was then sealed in a stainless-steel can and degassed at 400 °C in 0.1 Pa pressure. The canned powders were subsequently consolidated by a hot extrusion technique at 1150 °C. After the extrusion, the consolidated K3-ODS steel was thermally treated at 1050 °C for 1 hour. One ODS steel sample was also annealed at 900 °C, 168 hours (one week) for a thermal stability study. The chemical composition (in wt.%) of the consolidated material is C: 0.08, Si: 0.033, Cr: 16, W: 1.82, Al: 4.59, Ti: 0.28, Y_2O_3: 0.368, and Fe: balance [9]. Thin foils for TEM analysis were prepared by a standard procedure that includes slicing, grinding, and polishing the recovered fragments with the foils surface approximately perpendicular to the loading axis. Final thinning of the foils was performed using a standard twin-jet electropolishing technique in an electrolyte (90 vol.% acetic acid, 10 vol.% perchloric acid) at 30 V and room temperature. Microstructural characterization was performed using Phillips CM300 field-emission transmission electron microscope (accelerating voltage of 300 kV). A software package CaRIne Crystallography 3.1 was used to simulate electron diffraction patterns in order to identify the crystal structures of ODS nanoparticles.

RESULTS AND DISCUSSION

Typical microstructures of K3-ODS steel are shown in Figs. 1a and 1b. Elongated grains (Fig. 1a) and dense oxide nanoparticles (Fig. 1b), mainly $Y_4Al_2O_9$ (YAM) complex oxide, were observed. The oxide nanoparticle sizes in K3-ODS (Fig. 1b) are mostly ranging between 1.7 nm and 30 nm, with a mean particle size of 5.91 nm, and a particle density of 1.33 x 10^{22} m^{-3}. Here an orientation relationship between $Y_4Al_2O_9$ oxide phase and the matrix can be derived from the selected-area diffraction pattern: $(0\bar{1}1)_\alpha \parallel (2\bar{4}2)_{YAM}$ and $[011]_\alpha \parallel [432]_{YAM}$. The formation of $Y_4Al_2O_9$ oxide particles was identified and confirmed by matching several observed and simulated diffraction patterns of different zone axes, and an example is shown in Fig. 2 for the $[432]_{YAM}$-zone pattern. $Y_4Al_2O_9$ has a monoclinic structure and space group: $P2_1/c$ with a = 0.7375 nm, b = 1.0507 nm, c = 1.1113 nm, and β = 108.58° [11, 12], which is illustrated in Fig. 2c. By comparing the $Y_4Al_2O_9$ nanoparticles formed in the ODS steel with the starting Y_2O_3 particles (space group: Ia_3, a cubic structure with a_o = 1.06 nm [12], particle size: 15-30 nm [13]) used to fabricate the ODS steel, one can realize that the formation of oxide nanoparticles in ODS steels does not take place solely through a fragmentation mechanism. The formation of $Y_4Al_2O_9$ oxide reveals the occurrence of an internal oxidation reaction: $2Y_2O_3 + Al_2O_3 \rightarrow Y_4Al_2O_9$ during consolidation, which is governed by the oxygen affinity of alloying elements. That is, the formation of Y-Al complex oxides becomes predominant when both Al and Ti are present in ODS steels, which is in agreement with a conclusion made by Kasada et al [3].

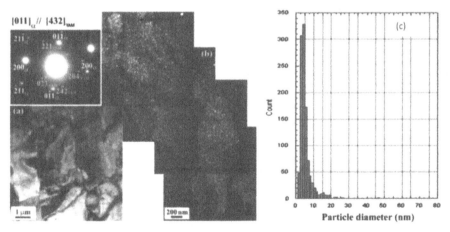

Fig. 1. (a) Bright-field TEM image shows typical grain morphology of K3-ODS steel, (b) dark-field TEM image and selected-area diffraction pattern of the $[011]_{Fe-Cr(\alpha)} \parallel [432]_{YAM}$-zone show the formation of dense $Y_4Al_2O_9$ nanoparticles in K3-ODS steel. (c) Size distribution of oxide nanoparticles measured from micrographs obtained using energy-filtered microscopy.

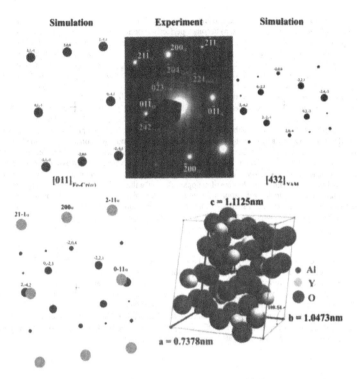

Simulation   Experiment   Simulation

$[011]_{Fe\text{-}Cr(\alpha)}$

$[432]_{YAM}$

$c = 1.1125nm$

$b = 1.0473nm$

$a = 0.7378nm$

● Al
● Y
● O

Fig. 2. (a) Observed and simulated diffraction patterns of the $[011]_{Fe\text{-}Cr(\alpha)}$-zone and the $[432]_{YAM}$-zone for identifying the formation of $Y_4Al_2O_9$ complex-oxide nanoparticles; (b) the overlap of simulated $[011]_{Fe\text{-}Cr(\alpha)}$-zone and $[432]_{YAM}$-zone diffraction patterns; (c) the crystal structure of $Y_4Al_2O_9$ complex-oxide compound .

Since the majority of oxide nanoparticles observed in K3-ODS steel has a diameter ranging between 2 nm and 30 nm, less attention was paid to nanoparticles larger than 30 nm. HRTEM images of a large (> 20 nm) and a small (< 10 nm) $Y_4Al_2O_9$ nanoparticles in K3-ODS steel sample are shown in Figs. 3 and 4, respectively. The large nanoparticle (viewing from the $[100]_{YAM}$ direction) tends to be nearly spherical in shape and is incoherent with the matrix. In addition, one can distinguish the core region with an appearance of lattice fringes from the outer shell region with a featureless appearance. The images of the large nanoparticle taken from two different defocus conditions seem to indicate that the lower portion of the large nanoparticle contains a disordered (or amorphous) domain that appears no clear lattice fringes as those appeared on the upper portion of the nanoparticle. In other words, the large nanoparticle is in fact not fully crystallized. On the other hand, ledges and facets can be seen at the interface between the small nanoparticle and the matrix, as shown in Figs. 4a and 4b, which indicates that the small nanoparticle tends to be coherent or semi-coherent with the matrix. In addition, small featureless domains can be readily found at the oxide/matrix interface shown in Fig. 4a. Here an orientation relationship between the $Y_4Al_2O_9$ oxide and the matrix, which is different from that in Figs. 1, can be derived from the fast Fourier transformation (FFT) image shown in Fig. 4: $(0\bar{1}1)_\alpha \parallel (\bar{1}\bar{1}5)_{YAM}$ and $[011]_\alpha \parallel [732]_{YAM}$. Figure 5 may shed some light on why featureless domains at the oxide/matrix interfaces form. Here nucleation of a crystalline $Y_4Al_2O_9$ domain (2 x 2 nm in size) can be clearly seen within a partially crystallized nanoparticle. The orientation relationship between the crystalline $Y_4Al_2O_9$ domain and the matrix, which can be derived from the FFT image: $(0\bar{1}1)_\alpha \parallel (2\bar{4}2)_{YAM}$ and $[011]_\alpha \parallel [432]_{YAM}$, is the same as that in Figs. 1. In general, facets, ledges, interface dislocations, and small amorphous domains are common features observed from the two interfaces shown in Figs. 4a and 5b. The two orientation relationships between the oxide nanoparticle and the matrix reveal that the {011} planes of the matrix act as habit planes for the nucleation of $Y_4Al_2O_9$ nanoparticles; the faceted and ledged oxide/matrix interfaces indicate the growth of small nanoparticles through a ledge mechanism.

13

Figure 6 shows the result of an ODS steel sample annealed at 900 °C for 168 hours. Here a small nanoparticle (< 10 nm) remains faceted and a large nanoparticle (> 20 nm) becomes perfectly spherical and fully crystallized without a core/shell structure. The observation of oxide nanoparticle without a core/shell structure after prolonged annealing seems to suggest that the core/shell structures of oxide nanoparticles formed in the as-fabricated ODS steels are far from chemical equilibrium. a revised three-stage formation mechanism of oxide nanoparticles is accordingly proposed based on the above HRTEM observations: (1) Fragmentation of starting Y_2O_3 particles to form finely-dispersed (nano or sub-nano) fragments during ball milling; (2) Agglomeration and amorphization of fragments mixed with matrix material to form clusters and agglomerates (designated as [MYO], M: alloying elements) during ball milling; (3) Crystallization of the amorphous oxide agglomerates to form oxide nanoparticles with a complex-oxide core and a solute-enriched (M') shell. The contents of complex-oxide core and solute-enriched shell are dependent on the compositions of different ODS steels. Y-Al complex-oxide ($Y_xAl_yO_z$) cores can form in Al-contained ODS steels such as the currently studied 16Cr-ODS, Y-Ti complex-oxide ($Y_xTi_yO_z$) cores can form in Ti-contained ODS steels with no addition of Al such as MA957 steel [7], and Y_2O_3 cores can form in ODS steels with no additions of Al and Ti such as Eurofer 9Cr-ODS steel [8]. The solute-enriched shells can be perceived as a result of the depletion of the solutes that are not involved in the internal oxidation reactions for the complex-oxide cores such as Cr-enriched shells in the nanoparticles of MA957 and Mn- and V-enriched shells in the nanoparticles of Eurofer 9Cr-ODS steel [7, 8]. The shell thickness is dependent on the size of nanoparticles since the larger the particle the more matrix material will participate in the agglomeration and amorphization stage and thus more solutes will be depleted from the oxide core during the crystallization stage. A solute-enriched shell forms when solute depletion rate from the core is greater than solute diffusion rate from the oxide/matrix interface during the crystallization stage.

SUMMARY

The oxide nanoparticles formed in 16Cr-4Al-2W-0.3Ti-0.3Y_2O_3 (K3) ODS steel are mainly $Y_4Al_2O_9$ (YAM) with a monoclinic structure. Large nanoparticles (> 20 nm) have a spherical shape and tend to be incoherent with the matrix; small nanoparticles (< 10 nm) accompany with facets and ledges at the oxide/matrix interfaces and tend to be coherent or semi-coherent with the matrix. A structure of crystalline oxide core in association with amorphous shell was observed in both large and small nanoparticles in as-fabricated K3-ODS steel. The core/shell structure becomes vanished after prolonged annealing at 900 °C for 168 hours, which suggests that the core/shell structure of oxide nanoparticles are far from chemical equilibrium. A three-stage formation mechanism of ODS nanoparticles is accordingly proposed to rationalize the core/shell structures of nanoparticles observed in as-fabricated ODS steels.

ACKNOWLEDGEMENTS

This work was performed under the auspices of the U.S. Department of Energy by Lawrence Livermore National Laboratory under Contract DE-AC52-07NA27344. Work at LLNL was funded by the Laboratory Directed Research and Development Program at LLNL under project tracking code 09-SI-003. The authors gratefully acknowledge Nick E. Teslich and Rick J. Gross for their efforts on TEM sample preparation.

REFERENCES

1. K. Ehrlich, Phil. Trans. R. Soc. Lond. A 357 (1999) 595.
2. S. Ukai, M. Fujiwara, J. Nucl. Mater. 307-311 (2002) 749.
3. R. Kasada, N. Toda, K. Yutani, H.S. Cho, H. Kishimoto, A. Kimura, J. Nucl. Mater. 341 (2005) 103.
4. T. Okuda, M. Fujiwara, J. Mater. Sci. Lett. 14 (1995) 1600.
5. Y. Kimura, S. Takaki, S. Suejima, R. Uemori, H. Tamehiro, ISIJ International, 39 (2) (1999) 176.
6. H. Sakasegawa, M. Tamura, S. Ohtsuka, S. Ukai, H. Tanigawa, A. Kohyama, M. Fujiwara, J. Alloys & Compounds 452 (2008) 2.
7. E. A. Marquis, Appl. Phys. Lett. 93 (2008) 181904.
8. M. Klimiankou, R. Lindau, A. Möslang, J. Nucl. Mater. 386-388 (2009) 553.
9. K. Yutani, H. Kishimoto, R. Kasada, A. Kimura, J. Nucl. Mater. 367-370 (2007) 423.
10. S. Uki, T. Nishida, H. Okada, T. Okuda, M. Fujiwara, K. Asabe, J. Nucl. Sci. Technol. 34 (3) (1997) 256.
11. A. Nørlund Christensen and R.G. Hazell, Acta Chemica Scandinavica 45 (1991) 226.
12. W.Y. Ching, Y.N. Xu, Physical Review B 59 (20) (1999) 12815.
13. V. de Castro, T. Leguey, M.A. Monge, A. Munoz, R. Pareja, D.R. Amador, J.M. Torralba, M. Victoria, J. Nucl. Mater. 322 (2003) 228.

14

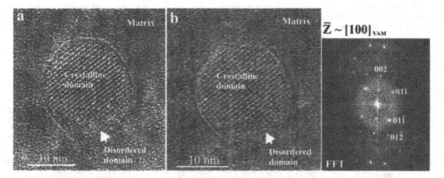

Fig. 3. HRTEM images of a large Y₄Al₂O₉ (YAM) nanoparticle at two different defocus conditions (a) -75 nm, (b) -50 nm.

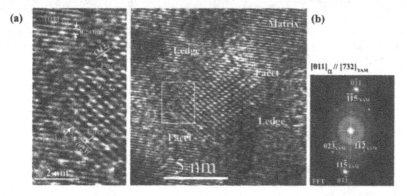

Fig. 4. HRTEM images of a small Y₄Al₂O₉ (YAM) nanoparticles; (a) The oxide/matrix interfacial structure observed from the marked area shown in (b). Facets, ledges, dislocations, and thin layer of featureless domains (marked by red arrows) can be readily seen at the interfaces. The orientation relationship between the nanoparticle and the matrix can be derived from the FFT image.

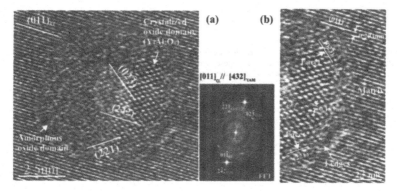

Fig. 5. HRTEM images show (a) the nucleation of a crystalline $Y_4Al_2O_9$ domain (2 x 5 nm) in an amorphous oxide particle; (b) facets, ledges, and amorphous remnant (marked by red arrows) can be seen at the oxide/matrix interface (indicated by a red arrow in (a)). Orientation relationship between the $Y_4Al_2O_9$ crystalline domain and the matrix can be derived from the FFT image.

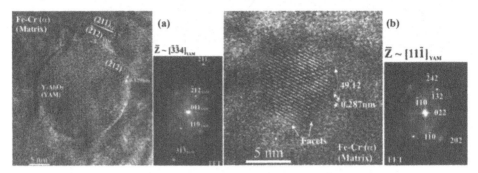

Fig. 6. HRTEM images of (a) a large $Y_4Al_2O_9$ nanoparticle (> 20 nm) and (b) a small nanoparticle (< 10 nm) after prolonged annealing at 900 °C for 168 hours. Notice that the small nanoparticle remains faceted and the large nanoparticle becomes perfectly spherical without a core/shell structure.

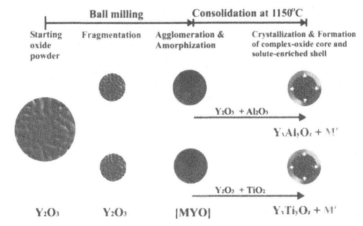

Fig. 7. Schematics of the proposed three-stage mechanism for the formation of oxide nanoparticles containing a core/shell structure during a mechanical alloying process; a solute-enriched shell forms when solute depletion rate at the core is greater than solute diffusion rate at the oxide/matrix interface during the crystallization stage.

16

Ceramic Materials and Wasteforms I

Mater. Res. Soc. Symp. Proc. Vol. 1215 © 2010 Materials Research Society   1215-V04-01

Mechanisms of Radiation Damage and Properties of Nuclear Materials

Gregory R. Lumpkin[1], Katherine L. Smith[1], Karl R. Whittle[1], Bronwyn S. Thomas[1], and Nigel A. Marks[2]

[1] Australian Nuclear Science and Technology Organisation, Menai, NSW, Australia
[2] Nanochemistry Research Institute, Curtin University of Technology, Perth, WA, Australia

ABSTRACT

Radiation damage effects in ceramics, e.g., nuclear waste forms, transmutation targets, and inert matrix fuels, may have important implications for the physical and chemical stability of these materials as the cumulative radiation dose increases over time. A key aspect of scientific research in this area is the ability to understand the fundamental damage mechanisms through the combination of experimental and atomistic modelling techniques. In this paper, we review some of the lessons learned from the significant body of data now available for pyrochlore-defect fluorite based materials, followed by an illustration of the advantages of working on simple compounds with well established interatomic potentials. We conclude the paper with a description of radiation damage processes in the $La_xSr_{1-1.5x}TiO_3$ defect perovskites, a system that includes phase transformations, short-range order effects, and complex defect behavior.

INTRODUCTION

Ceramic waste forms have been considered for many years to be viable alternatives for the disposal of nuclear waste in geological repositories. A major issue for waste forms is their behavior in response to alpha-decay damage and the presence of aqueous fluids for time scales on the order of 10^2 years for short-lived radionuclides (e.g., ^{90}Sr, ^{137}Cs) up to 10^{3-6} years or more for long-lived fission products and minor actinides. Radiation damage effects in these materials are studied using minerals from various geological systems, synthetic materials doped with short-lived (e.g., ^{238}Pu, ^{244}Cm) or medium-lived actinides (e.g., ^{239}Pu), or samples subjected to controlled irradiation sources, including electrons, alpha particles, neutrons, and ions. In the latter case, both bulk materials and thin TEM specimens can be irradiated as a function of ion mass and energy. In this report, we review some of the available data obtained by ion irradiation of TEM samples of pyrochlore-defect fluorite compounds, the TiO_2 polymorphs, and the La-Sr defect perovskites. The purpose of this paper is to illustrate some of the potential advantages of combining the experimental work with atomistic simulations of the threshold displacement, defect formation, and migration energies of the constituent atoms present in these crystalline compounds in order to understand variations in the radiation tolerance.

EXPERIMENTAL PROCEDURES

All of the ion irradiation experiments discussed in this paper have been conducted by various groups on thin TEM samples with 0.8-1.5 MeV Kr ions at the IVEM-Tandem Facility or the (now decommissioned) HVEM-Tandem Facility at Argonne National Laboratory. Readers are referred to the cited literature for information relating to the specific details of these experiments and the relevant static and dynamic atomistic calculations and simulations.

PYROCHLORE-DEFECT FLUORITE COMPOUNDS

Pyrochlore, defect fluorite, and related oxides are important technological materials with interesting magnetic, electrical, and fast ion transport properties. They have also been studied extensively as potential nuclear materials, e.g., as waste forms for actinides and fission products and as transmutation targets. Radiation damage effects due to alpha decay of Th and U were first studied in natural pyrochlores, where it is found that various compositions in the multi-component system (greatly simplified here) $NaCaNb_2O_6F-NaCaTa_2O_6F-CaUTi_2O_7-CaThTi_2O_7-Ln_2Ti_2O_7$ become amorphous with critical doses (D_c) of 10^{15} to 10^{16} alphas mg^{-1} or higher [1]. The critical dose appears to increase as a function of the geological age, possibly due to long-term diffusion or storage at elevated temperature in the Earth's crust. Actinide doping experiments [2] on $Gd_2Ti_2O_7$ illustrate that the total volume expansion is on the order of 5% (Figure 1). In synthetic systems, stoichiometric, $A_2B_2O_7$ III-IV pyrochlores exhibit a more or less systematic response to ion irradiation, ranging from complete amorphization to retention of crystallinity as a function of composition and the mass and energy of the impacting ions (typically noble gases with E = 0.5-1.5 MeV). For irradiations conducted with 1.0 MeV Kr ions, it has been shown that radiation "tolerance", e.g., resistance to structural damage as defined by the critical temperature for amorphization (T_c), is related to structural parameters, ionic radius ratios, and bonding criteria [3]. Atomistic simulations also demonstrate that there is a fundamental relationship with the energetics of cation and anion disorder in these compounds [4]. In the III-IV pyrochlore structure-field map defined by the radii of the ^{VIII}A-site and ^{VI}B-site cations we find decreasing T_c (increasing radiation tolerance) as a function of decreasing r_A/r_B. As a result, most lanthanide based pyrochlores with B = Ti are susceptible to amorphization with T_c values in the range of 480-1120 K whereas pyrochlore and defect fluorite compounds with B = Hf and Zr exhibit maximum T_c values of ~ 570 and 350 K, respectively.

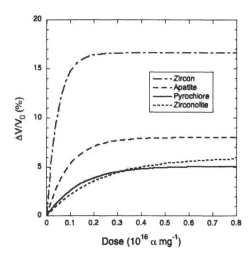

Figure 1. Comparison of total volume expansion for actinide doped materials. Synthetic zircon is doped with ^{238}Pu, whereas synthetic apatite, pyrochlore, and zirconolite are doped with ^{244}Cm. The very high expansion of zircon is undesirable from the viewpoint of cracking and enhanced dissolution in aqueous fluids. Pyrochlore exhibits the lowest volume expansion of any material known to become amorphous; the total volume expansion is similar to the unit cell expansion component of zircon. Unit cell expansion in pyrochlore is not well constrained, but is approximately half the total volume expansion.

It has been demonstrated that the radiation response (critical temperature) of pyrochlore can be modeled empirically based on the criteria listed above. In an effort to test the models, we have recently irradiated synthetic $Y_2Ti_{2-x}Sn_xO_7$ pyrochlores with x = 0.4, 0.8, 1.2, and 1.6, together with $Nd_2Zr_2O_7$, $Nd_2Zr_{1.2}Ti_{0.8}O_7$ and $La_{1.6}Y_{0.4}Hf_2O_7$. Determination of the critical amorphization fluence (F_c) as a function of temperature has revealed a dramatic increase in radiation tolerance with increasing Sn content on the pyrochlore B site. Non-linear least squares analysis of the fluence-temperature curves gave critical temperatures of 666 ± 4 K, 335 ± 12 K, and 251 ± 51 K for the samples with x = 0.4, 0.8, and 1.2, respectively. The sample with x = 1.6 appears to disorder to a defect fluorite structure at a fluence below 1.25×10^{15} ions cm^{-2} and remains crystalline to 5×10^{15} ions cm^{-2} at 50 K. Additionally, the critical fluence-temperature response curves were determined for $Nd_2Zr_{1.2}Ti_{0.8}O_7$ and $La_{1.6}Y_{0.4}Hf_2O_7$, and we obtained T_c values of 685 ± 53 K and 473 ± 52 K, respectively, for these pyrochlores. $Nd_2Zr_2O_7$ did not become amorphous after a fluence of 2.5×10^{15} ions cm^{-2} at 50 K, but there is evidence that it may amorphize at a higher fluence, with an estimated T_c of ~ 135 K. In the YTS pyrochlores, T_c appears to be linear in composition, and is globally linear in r_A/r_B and x(48f) for all samples investigated thus far [5].

The available results indicate that radiation tolerant pyrochlore-based nuclear waste forms with reduced volume expansion can be developed with suitable chemistry and properties, including the ability to incorporate impurities and neutron absorbers, together with a high level of resistance to attack by aqueous fluids. Furthermore, using the existing data and models as a guide, it may also be possible to develop advanced technological materials using pyrochlore compounds with certain electrical/magnetic properties locally modified by ion irradiation.

MODEL SYSTEMS: TiO$_2$ POLYMORPHS

More recently, we have returned to simple systems in an effort to separate the effects of structure and composition on the radiation tolerance of nuclear materials. Wang et al. [6] have already shown that this is a useful exercise in their work on the GeO$_2$ polymorphs. The results of our irradiation experiments conducted at 50 K on the samples with low levels of chemical impurities demonstrate that synthetic rutile remains crystalline up to a fluence of 5×10^{15} ions cm^{-2}, indicating that there is no significant accumulation of amorphous domains. In comparison, brookite and anatase become amorphous during 1.0 MeV Kr irradiation with F$_c$ values of 8.1 ± 1.8×10^{14} ions cm^{-2} and 2.3 ± 0.2 x 10^{14} ions cm^{-2}, respectively. Furthermore, irradiation at higher temperatures allowed us to determine T$_c$ = 170 K for brookite and 242 K for anatase. We have also examined two natural rutile samples from Graves Mountain, Georgia (sample GM), and from Alexander County, North Carolina (sample NC). Analytical data determined by SEM-EDX indicate that the GM rutile has about 1.7 wt% impurities (mainly Fe > V > Cr) whilst the NC rutile has about 1.2 wt% impurities (mainly Cr > V > Nb). Results of the irradiation experiments indicate that the GM and NC rutile samples become amorphous at critical fluences of 9.2 ± 0.7 x 10^{14} ions cm^{-2} and 8.6 ± 0.5 x 10^{14} ions cm^{-2}, respectively, at 50 K. The temperature response curve was determined for the GM rutile and gives a T$_c$ = 207 K. The fluence-temperature data are compared in Figure 2.

For samples with low levels of chemical substitution, we have shown that the radiation response data of rutile, brookite, and anatase correlate inversely with the number of shared edges per TiO$_6$ octahedron in each structure: two for rutile (tetragonal, site symmetry mmm), three for brookite (orthorhombic, site symmetry 1), and four for anatase (tetragonal, site symmetry

21

Σ4m2). A characteristic feature of edge sharing in octahedral framework structures is an increase in the distortion of the coordination polyhedra as the shared edges become shorter and the unshared edges become longer. As the number of shared edges increases, the structures become more distorted, evidenced by increasing octahedral edge length distortion, bond angle variance, and quadratic elongation from rutile to brookite to anatase (see Table 1). Similar distortions are characteristic of amorphous oxide materials. Note that mean bond lengths and bond length distortion do not necessarily correlate with other distortion parameters.

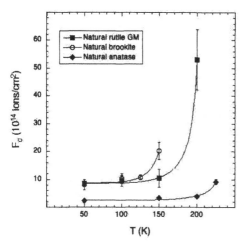

Figure 2. Comparison of the fluence-temperature response of natural rutile, brookite, and anatase. The natural rutile sample is from Graves Mountain, Georgia, and contains ~ 1.7 wt% impurities, including transition metals and water (as protons located near ½ ½ 0). Natural brookite and anatase contain < 0.5 wt% impurities (mainly transition metals). Note that high-purity synthetic rutile cannot be rendered amorphous up to a fluence of 50×10^{14} ions cm^{-2} at 50 K. These results suggest that structure and chemistry both play a role in the radiation response of the TiO_2 polymorphs.

Atomistic simulations reported by Thomas et al. [7] for rutile show that the E_d for O displacements ranged from 50 to 85 eV depending upon direction, with an average value of 65 eV for a defect formation probability of 50%. For Ti displacements, E_d ranged from 80 to 160 eV depending upon direction, with an average value of 130 eV. In contrast to O, the E_d for Ti is highest along $\langle 111 \rangle$ and lowest along $\langle 001 \rangle$ directions. The higher average value for Ti is related to a) a higher lattice binding energy, b) greater electrostatic attraction between the vacancy and interstitial, and c) PKA energy transfer to O atoms. We have conducted additional calculations for anatase, finding that the E_d for O ranges from 20 to 100 eV with an average of about 55 eV. Due to difficulties in the statistics and determination of directional behavior, only a minimum value of $E_d > 90$ eV has been determined for Ti in anatase. Similar difficulties were encountered for brookite, where we currently find E_d values to be > 60 eV for O and > 80 eV for Ti.

Thomas et al. [7] also calculated formation energies of 9-11 eV for O Frenkel defects in rutile. These are of the split, channel, and channel wall type, with the split defects being the most common type. The O migration energy is < 0.1 eV as estimated from static and dynamic calculations and the interstitials migrate from split interstitial to split interstitial via channel sites, enabling movement in three dimensions, but are particularly active in the (001) plane. The Ti Frenkel defect energy was found to be approximately 18 eV and the Ti interstitials prefer the channel sites. The energy barrier for Ti migration was estimated to be 0.3 eV. Migration occurs from channel site to channel site along ⟨001⟩ and also by split interstitials and coordinated movements of many Ti atoms in the (001) plane, e.g., along ⟨100⟩ directions.

Additional static defect calculations were reported in reference [8], where we observed that the O Frenkel defect formation energies are very similar for all three polymorphs, with values of 9.0, 9.6, and 8.0 eV for rutile, brookite, and anatase, respectively. With regard to defect configurations, a common feature of all three polymorphs is the occurrence of the split oxygen interstitial. For Ti Frenkel defects, results of the calculations give formation energies of 18.1, 21.4, and 7.3 eV for rutile, brookite, and anatase, respectively. In both cases, the defect formation energies scale in the order anatase < rutile < brookite. We have also undertaken some preliminary modelling of O migration and our results suggest that the barriers scale in the order rutile < anatase < brookite.

Table 1. Volume data, polyhedral distortion data, threshold displacement energies, defect formation energies, and defect migration energies for the TiO_2 polymorphs.

	Rutile	Brookite	Anatase
Shared edges	2	3	4
Molar volume, V_m ($\mathring{A}^3$ mol^{-1})	62.45	64.21	68.12
Octahedral volume, V_o ($\mathring{A}^3$)	9.903	9.761	9.473
Bond length distortion, Δ_{Ti-O}	0.790	2.754	1.063
Edge length distortion, Δ_{O-O}	3.097	4.752	6.912
Bond angle variance, σ^2	63.08	143.51	231.34
Ti TDE, E_d (eV)	130	> 80	> 90
O TDE, E_d (eV)	65	> 60	~ 55
Ti Frenkel formation energy (eV)	18.1	21.4	7.3
O Frenkel formation energy (eV)	9.0	9.6	8.0
Ti migration barrier energy (eV)	0.3	> Rutile ?	> Brookite ?
O migration barrier energy (eV)	< 0.1	> Rutile ?	> Brookite ?

In previous work [8,9], we performed molecular dynamics (MD) simulations of the evolution of local regions of damage in rutile, brookite, and anatase using a thermal spike approach. Our simulations reveal that the time required for peak development of the thermal spike and the maximum number of defects in the spike both follow the order rutile < brookite < anatase. Annealing of thermal spike damage occurs beyond a time frame of 0.1-0.2 ps. Rutile has completely annealed within 3 ps, brookite retains a few defects beyond 3 ps, and anatase retains many more defects out to 5 ps. In fact, the MD simulations reveal that a "stable" amorphous region is created in anatase that is slightly smaller that the intial spike dimension. Data analysis reported in reference [9] shows that, up to the peak level of spike damage, the numbers of O and Ti defects are close to the stoichiometric ratio of 2:1. Beyond the peak of damage, however, this is not the case and we see important differences in the dynamic defect behavior as O defects migrate back to lattice sites more quickly than Ti defects. Although the thermal spike model provides qualitative agreement with experimental data, additional MD simulations need to be performed in order to understand the damage produced by recoils as a function of energy.

COMPLEX SYSTEMS: La-Sr DEFECT PEROVSKITES

All perovskite samples in the $La_xSr_{1-1.5x}TiO_3$ system pass through the crystalline-amorphous transformation. However, as shown in Figure 3, the critical temperature for amorphization varies nonlinearly with composition, decreasing from 394 K for $SrTiO_3$ to a minimum of 275 K for $La_{0.2}Sr_{0.7}TiO_3$, followed by an increase with composition to 865 K for $La_{0.67}TiO_3$ [10]. With each increment of x in this system, there are 0.5x vacancies on the A-site of the perovskite structure. Such cation vacancies may themselves play an important role in determining the defect mobility. In combination with the work of previous authors [11,12], the electron microscopy results and IVEM data presented by Smith et al. [10] suggest that there is a vacancy-assisted recovery mechanism in the cubic perovskites with compositions with $x \leq 0.2$ that is not directly associated with the cubic-tetragonal phase transformation and that vacancy-enhanced amorphisation occurs for both cubic and orthorhombic perovskites with compositions in the range x= 0.2–0.67.

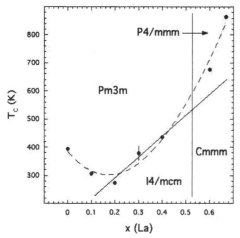

Figure 3. Summary of experimental data for the system $La_xSr_{1-1.5x}TiO_3$ showing phase boundaries and the T_c response for compositions with x = 0.0, 0.1, 0.2, 0.3, 0.4, 0.6, and 0.67. The dashed line serves as a guide for the T_c trend only. In the cubic phase field, these compounds become more radiation resistant up to x = 0.2. For x = 0.2-0.4 the data fall on or very near the tetragonal to cubic phase transformation boundary, suggesting that the critical temperature may be coupled with the phase transformation. For x > 0.4 the data fall within the field of the tetragonal layered structure. Short range order is present for $x \geq 0.3$ at room temperature in this system.

Thomas et al. [11] have shown that, up to La substitution of approximately x = 0.2, electrostatic effects in the cubic phase promote the association of La-vacancy pairs and the dissociation of vacancy-vacancy pairs. Although defect formation and migration energies in this range of x are currently unknown, atomistic simulations of $SrTiO_3$ [12] using density functional theory (DFT) suggest that the O Frenkel defect energy is lower than Sr and Ti Frenkel defect energies, and these are in turn lower than the Sr/Ti antisite formation energy (see Table 2). Defect migration energies have also been evaluated and suggest that migration barriers are lowest for O and Sr interstitials and O vacancies. For x = 0.0–0.2, T_c decreases from 394 to 275 K and we have proposed that this mechanism involves the electrostatic driving forces noted above together with enhanced cation-vacancy mobility due to the increasing concentration of intrinsic A-site vacancies.

In the composition range x = 0.2–0.4, T_c begins to increase as the driving force for vacancy-vacancy dissociation begins to change as strain forces increase and electrostatic forces decrease [10,11]. Ordered domains also begin to form and grow within this range of compositions. The data shown in Figure 3 suggest that in this composition range, the critical temperature for amorphization may be coupled with the ferroelastic tetragonal to cubic phase transformation. For samples with x > 0.4, the La-vacancy interactions and strain forces become dominant in the highly ordered crystals and T_c increases to 865 K in the La end-member. Apart from inversion of defect interactions, the mechanism for this decrease in radiation tolerance is currently unknown, as the defect energies and migration energies have not been determined for any composition in the system other than SrTiO$_3$. However, it is possible that the process of inversion and dominance of strain interactions for La rich compositions have altered the defect formation energies and/or migration energies. It is also possible that local chemical fluctuations might also affect the electronic behavior of these solid-solutions which are known to exhibit metal-insulator transitions, with further implications for defect mobility. This holds potential for future work.

Table 2. Threshold displacement energies, cation antisite (Sr/Ti) formation energies, Frenkel defect formation energies, and defect migration energies for SrTiO$_3$ [12].

	Sr/Ti	Sr	Ti	O
TDE, E_d (eV)	—	70	140	50
Defect formation energy (eV)	9.3	16.5	18.2	10.1
DFT defect formation energy (eV)	13.2	9.0	—	7.9
Vacancy migration energy (eV)	—	3.9	11	0.9
Interstitial migration energy (eV)	—	0.3	4.6	0.3

CONCLUSIONS

The radiation tolerance of III-IV A$_2$B$_2$O$_7$ pyrochlore-defect fluorite compounds is highly correlated with structural data, bonding parameters, and defect formation energies. Extensive atomistic modelling results indicate that A-B cation antisite and anion Frenkel defect formation energies are important in these compounds and are generally closely related mechanisms. Studies of the TiO$_2$ polymorphs indicate that polyhedral distortion and volume properties are important factors in determining the radiation tolerance. This is partly equivalent to saying that distorted structures provide alternative stable atomic configurations in damaged materials whereas volume strain may work against this by providing a driving force for damage recovery. Although the polymorphs have no inherent cation/anion disorder mechanism, rutile, in particular, has very low migration barriers leading to cooperative Ti-O defect recovery pathways. Finally, studies of the complex radiation behavior of perovskites in the system La$_x$Sr$_{1-1.5x}$TiO$_3$ reveal that electrostatic and strain effects are important in promoting La-vacancy association or dissociation, respectively. Lower migration barriers, driven in part by increasing A-site vacancies, may be responsible for increasing the radiation tolerance up to x = 0.2. At this point as the electrostatic and strain effects begin to invert, the critical temperature may couple with the phase transformation up to x = 0.4. Above, this point, as ordered domains grow beyond a critical size in the low temperature tetragonal phase, the critical temperature decouples from the phase transformation. For x values above 0.4, the critical temperature lies within the field of the high temperature ordered tetragonal phase.

ACKNOWLEDGMENTS

We are grateful to the staff of the IVEM-Tandem Facility at Argonne National Laboratory for assistance during the ion irradiation work. The IVEM-Tandem Facility is supported as a User Facility by the U.S. DOE, Basic Energy Sciences, under contract W-31-10-ENG-38. We acknowledge financial support from the Access to Major Research Facilities Programme (a component of the International Science Linkages Programme established under the Australian Government's innovation statement, Backing Australia's Ability). Part of this work was supported by the Cambridge-MIT Institute (CMI), British Nuclear Fuels Limited (BNFL), and EPSRC Grant (EP/C510259/1) to G.R. Lumpkin.

REFERENCES

1. G.R. Lumpkin and R.C. Ewing, *Phys. Chem. Minerals* **16**, 2-20 (1988).

2. W.J. Weber, J.W. Wald, and Hj. Matzke, Mater. Letters 3, 173-180 (1985).

3. G.R. Lumpkin, M. Pruneda, S. Rios, K.L. Smith, K. Trachenko, K.R. Whittle, and N.J. Zaluzec, *J. Solid State Chem.* **180**, 1512-1518 (2007).

4. L. Minervini, R.W. Grimes, and K.E. Sickafus, *J. Amer. Ceram. Soc.* **83**, 1873-1878 (2000).

5. G.R. Lumpkin, K.L. Smith, M.G. Blackford, K.R. Whittle, E.J. Harvey, S.A.T. Redfern, and N.J. Zaluzec, *Chem. Mater.* **21**, 2746-2754 (2009).

6. S.X. Wang, L.M. Wang, and R.C. Ewing, *Nucl. Instr. Meth. Phys. Res. B* **175-177**, 615-619 (2001).

7. B.S. Thomas, N. A. Marks, L. R. Corrales, and R. Devanathan, *Nucl. Instr. Meth. Phys. Res. B* **239**, 191 (2005).

8. G.R. Lumpkin, K.L. Smith, M.G. Blackford, B.S. Thomas, K.R. Whittle, N.A. Marks, and N.J. Zaluzec, *Phys. Rev. B* **77**, 214201 (2008).

9. N.A. Marks, B.S. Thomas, K.L. Smith, and G.R. Lumpkin, *Nucl. Instr. Meth. Phys. Res. B* **266**, 2665-2670 (2008).

10. K.L Smith, G.R. Lumpkin, M.G. Blackford, M. Colella, and N.J. Zaluzec, *J. Appl. Phys.* **103**, 083531 (2008).

11. B.S. Thomas, N. A. Marks, and P. Harrowell, *Phys. Rev. B* **74**, 214109 (2006).

12. B.S. Thomas, N. A. Marks, and B.D. Begg, *Nucl. Instr. Meth. Phys. Res. B* **254**, 211 (2007).

Modeling Complex Materials I

Mater. Res. Soc. Symp. Proc. Vol. 1215 © 2010 Materials Research Society   1215-V05-07

Multiscale Modeling of Helium-Vacancy Cluster Nucleation under Irradiation: a Kinetic Monte-Carlo Approach.

Tomoaki Suzudo , Masatake Yamaguchi, Hideo Kabraki and Ken-ichi Ebihara
Center for Computational Science and e-Systems, Japan Atomic Energy Agency, Tokai-mura, 319-1195, Japan

ABSTRACT

We applied ab initio calculation and an object kinetic Monte Carlo modeling to the study of He-vacancy cluster nucleation under irradiation in bcc and fcc Fe, which are surrogate materials for ferritic/martensitic and austenitic steels, respectively. The ab initio calculations provided parameters for the object kinetic Monte Carlo model, such as the migration energies of point defects and the dissociation energies of He and vacancy to He-vacancy clusters. We specially focused on the simulation of high He/dpa irradiation such as He-implantation into the materials and tracked the nucleation of clusters and the fate of point defects such as SIAs, vacancies, and He atoms. We found no major difference of He-vacancy cluster nucleation between bcc and fcc Fe when we ignore the intracascade clustering even if the migration energies of point defects are significantly different between the two crystals.

INTRODUCTION

He produced in steel by high-energy neutron radiation causes significant change in its mechanical properties, see e.g. [1] and references therein. The swelling caused by the nucleation and growth of He bubbles is a primary concern. Both of ferritic/martensitic and austenitic steels are considered as candidate structural materials of fast breeder reactors, and the swelling behavior of these steels are comparatively discussed [2, 3]. Many experimental evidences show that ferritic/martensitic steels are generally more swelling-resistant than austenitic steels [2]. This difference is numerically studied by molecular dynamics, and the defect configurations after a radiation cascade in bcc Fe and fcc Cu reveals that intracascade vacancy clusters in fcc Cu are larger than those produced in bcc Fe [4]. This result is considered as a cause of swelling vulnerability of fcc metals [5]. However, the other factors influencing the swelling behaviors, such as point-defect migrations and point-defect dissociation from clusters, have not been thoroughly examined yet.

In the present paper, we modeled He-vacancy clustering and conducted comparative studies between bcc and fcc Fe, which are surrogate materials for ferritic/martensitic and austenitic steels, respectively. Using ab initio methodology, we calculated the migration energies of point defects and He and vacancy binding energies to He-vacancy clusters, all of which are key parameters for the nucleation and growth of He-vacancy clusters. Then we conducted object kinetic Monte Carlo simulations using the parameters obtained by the ab initio calculations. The main aim of this paper is to compare He-vacancy clustering behavior between bcc and fcc Fe under irradiation.

AB INITIO CALCULATIONS

We derived ab initio energies related to the nucleation of He-vacancy clusters under irradiation. The unit cells used for the modeling are shown in figure 1. Atomic structure with a specified defect was relaxed using Vienna Ab initio Simulation Package (VASP) with Projector Argumented Wave potential [6,7]. In this calculation we used the cutoff energy of 280 eV for the plane wave basis, the Monkhorst-Pack 3x3x3 k-point mesh, and the Meshfessel-Paxton smearing with 0.1 eV width. Only for the cases of bcc Fe, spin-polarized calculations were performed.

First, we calculated migration energies of point defects such as vacancy, He atom at tetrahedral interstitial sites, and dumbbell-type self-interstitial atom (SIA), using the Nudged Elastic Band method. Table I shows the results.

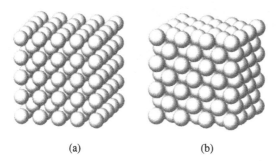

(a)   (b)

Figure 1. Unit cells used for ab-initio calculations (a) bcc Fe 4x4x4 cell with 128 Fe atoms, and (b) fcc Fe 3x3x3 cell with 108 atoms.

Table I. Migration energies of point defects by ab-intio method

	In bcc Fe	In fcc Fe
Vacancy	0.67	1.36
Interstitial He	0.06	0.24
SIA	0.32	0.32

Notice that the migration energy of interstitial He atoms in bcc Fe is extremely small, suggesting that He can migrate athermally. The vacancy migration energy in bcc Fe is also significantly smaller than that in fcc Fe, suggesting that vacancies in fcc Fe survive for longer periods than those in bcc Fe.

Second, we calculated He and vacancy binding energies to various He-vacancy clusters ($He_n V_m$: n and m =0 to 4). For bcc Fe, similar calculations already exist [8], but to make the comparison between bcc and fcc Fe clearer we recalculated the cases for bcc Fe. The results are summarized in figure 2(a), and it indicates that for large He-to-vacancy ratio vacancy binding energies for fcc Fe are meaningfully larger than those for bcc Fe. In addition, He binding energies for fcc Fe seem slightly larger than those for bcc Fe. We also confirmed the results for bcc Fe are not significantly different from the values in Ref. [8]. The dissociation energies were calculated as summation of the binding energy and the migration energy in Table I, see figure 2(b). Note that the dissociation curves of fcc Fe are slightly larger than those of bcc Fe; this means He-vacancy cluster in fcc Fe are slightly more stable than those in bcc Fe. However, this does not necessarily mean that He-vacancy clusters can grow more easily in fcc Fe as described in the following section that considers the kinetics of the nucleation and growth.

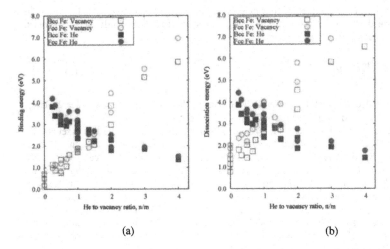

(a)   (b)

Figure 2. (a) Binding energy and (b) dissociation energy of vacancy and He atom to He-vacancy clusters in bcc and fcc Fe.

OBJECT KINETIC MONTE CARLO CALCULATIONS
Modeling method

Because we used a typical method of object kinetic Monte Carlo simulation, as reported elsewhere [9-12], only distinctive part of the modeling is described in the following. Note that calculation conditions given below are the same in both bcc and fcc Fe unless otherwise noted.

We considered the evolution of lattice defects caused by continuous radiation into the materials of $(100nm)^3$ cube. The radiation was simulated by random insertion of He atoms and Frenkel pairs. We ignored intracascade clustering and concentrated on clustering by slow migrations of defects after radiation. Helium atoms were inserted at tetrahedral interstitial sites at a constant rate according to the given He/dpa ratio.

Helium atoms migrate through tetrahedral interstitial sites until they get trapped by a vacancy, another He atom, or He-vacancy (or pure vacancy) cluster. Self-interstitial atoms (SIAs) lie near a lattice site forming a dumbbell with a Fe atom at the lattice site, and they migrate between two neighboring lattice sites. Helium atoms, SIAs, and vacancies jump with the given probability of

$$P_m = (1/6\delta)\ D_m^{\ 0}exp\{-E_m/kT\}, \qquad\qquad (1)$$

where δ is the migration jump distance that are determined by the crystal structures; $D_m{}^0$ the pre-exponential factor for diffusion constant; E_m the migration energy; k the Boltzmann constant; T the absolute temperature. Note that δ, $D_m{}^0$, E_m are dependent on the type of point defect. For E_m we used the values in Table I. The $D_m{}^0$'s values of vacancy, interstitial He and SIA were set at 1.15×10^{-2} cm^2/s [9], 1.7×10^{-3} cm^2/s [12] and 2.09×10^{-3} cm^2/s [9], respectively; identical $D_m{}^0$s were used for both bcc and fcc Fe.

Each defect is associated with its inherent reaction sphere whose center is located at the position of the defect, and if spheres of any two defects overlap they instantly merge and react. The radius of each defect R_r is given by

$$R_r = \sqrt[3]{\frac{3m\Omega}{4\pi}} + R_0, \qquad (2)$$

where m is the size of defect; for example, 1 for vacancy, SIA and interstitial He clusters He$_n$, n for SIA clusters I_n and vacancy clusters V_n, and m for He-vacancy clusters He$_n$V$_m$; Ω is the atomic volume of Fe and is set at 1.19×10^{-23} cm^3; R_0 is an offset of the radius and is set at 0.125 nm. The reaction between an SIA and a vacancy cause recombination; two SIAs cause I_2; I_n and I_m cause I_{n+m}. Similarly He$_n$V$_m$ ($n,m=0,1,2,...$) and He$_{n'}$V$_{m'}$ ($n',m'=0,1,2,...$) causes He$_{n+n'}$ V$_{m+m'}$, where He$_n$V$_0$ and He$_0$V$_n$ mean He$_n$ and V$_n$, respectively. There is no size limit of clusters except that tetrahedral site can hold four He atoms at most, i.e. He$_n$ ($n=1,2,3,4$). Besides, any He$_n$V$_m$s with $n/m>4$ are not allowed. As you see in the next section, this condition is not important because n/m is generally less than 2 for both in bcc and fcc Fe. All clusters are immobile, except that He$_1$V$_1$ migrates with the same probability of vacancy [12]. To simulate dislocations working as sinks for point defects, we randomly inserted immobile point-wise sinks in the simulation box. The concentration of these sinks is 10^{-6} appm [13].

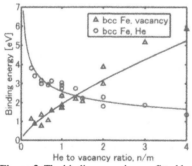

Figure 3. The binding energies are fitted by exponential functions.

We ignored SIA dissociation from SIA clusters, but we considered the dissociation of a He atom and vacancy from He-vacancy clusters with the given probability P_d, that is given by

$$P_d = (1/6\delta) \, D_m{}^0 exp\{-(E_m + E_b)/kT\}, \qquad (3)$$

where E_b is the binding energy given in figure 2. For larger clusters ($n>4$ or $m>4$), we used the following equations for bcc and fcc Fe;

$$E_{b,V}^{BCC} = 1.92 \, \lambda^{0.723} \tag{4}$$
$$E_{b,He}^{BCC} = 2.59 \, \lambda^{-0.323} \tag{5}$$
$$E_{b,V}^{FCC} = 2.12 \, \lambda^{0.770} \tag{6}$$
$$E_{b,He}^{FCC} = 2.84 \, \lambda^{-0.333} \tag{7}$$

The subscripts of V and He denote vacancy and He, respectively; λ is the He-to-vacancy ratio of a cluster. These equations were derived by fitting the numerical data as seen figure 3.

Results

We conducted the calculation up to ~0.55×10^{-3} dpa with 4×10^{-7} dpa/s and 5000 He-appm/dpa for both bcc and fcc Fe. The temperature was set at 823K. With these conditions, we simulated He implantation experiments both to ferritic and austenitic steels by Hasegawa et al.[14]. In this experiment they find that cavities emerge in the ferritic steel with lower dose, which is opposite to in-reactor neutron irradiation. In the both cases of crystal, a few He-vacancy clusters (He_nV_m) were nucleated. The sizes of He_nV_m clusters are shown in n-m space (figure 4). It is difficult to draw clear conclusions on the difference between bcc and fcc Fe because of the poor statistics, but at least one can say that drastic differences are not seen in spite of the significant difference in the migration energies of vacancy and helium atom as shown in Table I. Besides, this result did not qualitatively match the He-implantation experiments [14]. The size distributions of SIA clusters of bcc and fcc Fe are also shown in figure 5, where we did not find major difference between bcc and fcc Fe either. The result here suggests that the nucleation and growth rates of He-vacancy cluster in bcc and fcc Fe are similar. As seen in Table I, the migration energies of point defects are different between bcc and fcc Fe, and these differences do not seem to influence the long-time behavior. Note that the whole results could change if we conduct the simulations with taking intracascade clustering into consideration; this study is in progress.

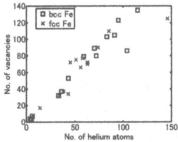

Figure 4. The distribution of He_nV_m clusters after the irradiation of ~0.55×10^{-3} dpa. Five cases of simulations are accumulated to gain better statistics.

CONCLUSIONS

He-vacancy clustering in bcc and fcc Fe was numerically studied by ab initio and object kinetic Monte Carlo calculations. The ab initio calculations indicated that the migration energies of interstitial He atom and vacancies in bcc Fe are significantly smaller than those in fcc Fe. The He and vacancy binding energies to He_nV_m were also calculated by ab initio method. These results were reflected in the numerical simulation of the He-vacancy cluster nucleation and growth using an object kinetic Monte Carlo method. We observed He-vacancy clusters nucleated both in bcc and fcc Fe and found no major difference of clustering behaviors between the two crystals even if the migration energies of point defects are significantly different.

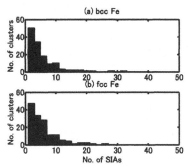

Figure 5. Size distribution of SIA clusters after the irradiation of 0.55×10^{-3} dpa.

ACKNOWLEDGMENTS

The authors wish to acknowledge fruitful discussion with Dr. E. Wakai, Dr. K. Aoto of Japan Atomic Energy Agency and Prof. A. Hasegawa of Tohoku University Japan. The present study includes the result of "R&D Project on Irradiation Damage Management Technology for Structural Materials of Long-life Nuclear Plant" entrusted to Japan Atomic Energy Agency by the Ministry of Education, Culture, Sports, Science and Technology of Japan (MEXT).

REFERENCES

1. H. Trinkaus, B.N. Singh, *J. Nucl. Mater.* **323,** 229(2003).
2. F.A. Garner, *Materials Science and Technology*, **Vol. 10A, Chapter 6**, VCH, Germany (1994).
3. F.A. Garner, M.B. Toloczko, B.H. Sencer, *J. Nucl. Mater.* **276**, 123(2000).
4. S.J. Zinkle, *Phys. Plasmas*, **12**, 0581101(2005).
5. M.J. Caturla, T. Diaz de la Rubia, M. Fluss, *J. Nucl. Mater.* **323**, 163(2003).
6. G. Kresse, J. Furthmüllar, *Phys. Rev. B* **47**, R558 (1993).

7. G. Kresse, D. Joubert, *Phys. Rev. B* **59**, 1758 (1999).
8. C.-C. Fu, F. Willaime, *Phys. Rev. B* **72**, 064117(2005).
9. M.J. Caturla, N. Soneda, E. Alonso, B.D. Wirth, T. Diaz de la Rubia, J.M. Perlado, *J. Nucl. Mater.* **276**, 13(2000).
10. C. Domain, C.S. Becquart, L. Malerba, *J. Nucl. Mater.* **335**, 121 (2004).
11. A.F. Voter, *Radiation Effects in Solids* **235**, 1 (2007).
12. C.S. Deo et al., *J. Nucl. Mater.* **361** 141(2007).
13. J. Rotller, D.J. Srollovitz, R. Car, *Phys. Rev. B*, **71**, 064109(2005).
14. A. Hasegawa, S. Nogami, Y. Sato, private communication.

Metallic Materials I

Mater. Res. Soc. Symp. Proc. Vol. 1215 © 2010 Materials Research Society   1215-V06-02

Radiation Response of a 12YWT Nanostructured Ferritic Steel

M. K. Miller, D. T. Hoelzer and K. F. Russell
Materials Science and Technology Division, Oak Ridge National Laboratory,
Oak Ridge, TN 37831-6136, USA

ABSTRACT

In order to evaluate the radiation response of 12YWT nanostructured ferritic steel to high dose neutron irradiation, the solute distribution, and size, number density, and compositions of nanoclusters in the unirradiated condition and after neutron irradiation to a dose of 3 dpa at a controlled temperature of 600 °C were estimated by atom probe tomography. No statistical difference in the average size or size distribution of the nanoclusters was found between the unirradiated and irradiated conditions. Therefore, these nanostructured ferritic steels are promising candidate materials for use under extreme conditions in future generations of advanced reactors.

INTRODUCTION

Iron alloys have played an important role in key components of commercial nuclear reactors. For example, the pressure vessels of nuclear reactors are fabricated from a variety of pressure vessel steels, such as A533B, A302B and their Russian counterparts. These nuclear reactors were designed to operate for a minimum of 20 years at an operating temperature of less than 350 °C. New generations of power generating systems are needed to meet future energy demands, which will require new materials with high tolerance to extreme environments - typically end-of-life neutron doses of up to 200 displacements per atoms (dpa) at temperatures of up to at least 750 °C.

Although iron-based alloys are typically not considered for long-term use at high temperatures due to significant grain growth and coarsening of precipitates, one class of ferritic steels that is under consideration for these extreme environments is the nanostructured ferritic steels - formerly referred to as oxide dispersion strengthened (ODS) steels. Nanostructured ferritic steels, such as 9Cr, 12YWT, 14YWT and MA957 alloys are produced by mechanically alloying pre-alloyed metals and yttria powders. This fabrication method forces all the elements in the powders into solid solution [1] and produces a high concentration of vacancies, thereby creating a new class of materials with remarkable properties.

Atom probe tomography (APT) characterizations have demonstrated that there are high number densities of titanium-, oxygen- and yttrium-enriched nanoclusters in these nanostructured ferritic steels [2-4]. The nanoclusters and the associated fine grain size are stable during high temperature isothermal aging at temperatures up to 1400 °C [3] and during long term creep at elevated temperatures (850 °C) [4]. Consequently, these unique materials are candidates for use under extreme conditions in future generations of advanced reactors.

Atomic displacement cascades produced during neutron or ion irradiation may induce mechanisms that could potentially destabilize or destroy the nanoclusters, change the vacancy and interstitial atom distribution, and thereby change the properties. Increasing the vacancy concentration may potentially enhance diffusion, which may result in a coarsening of the nanoclusters with a change in their number density. However, preliminary atom probe tomography characterizations have demonstrated that the nanoclusters in these alloys were not changed significantly after exposure to high doses of ion or proton irradiation [5,6].

In this study, the solute distribution associated with, and the stability of, the nanoclusters in a 12YWT nanostructured ferritic steel after exposure to neutron dose irradiation have been investigated by atom probe tomography.

EXPERIMENTAL

The composition of the mechanically alloyed 12YWT nanostructured ferrite steel is Fe-12.3 wt% Cr, 3% W, 0.39% Ti, + 0.25% Y_2O_3 [Fe-13.3 at. % Cr, 0.92% W, 0.46% Ti, 0.13% Y and 0.19% O]. All concentrations quoted in the remainder of this paper are given in atomic percent. Details of the preparation of these mechanically alloyed materials have been reported previously [2]. The 12YWT nanostructured ferritic steel was characterized in the unirradiated condition and after neutron irradiation in Oak Ridge National Laboratory's High Flux Isotope Reactor to a dose of 3 dpa at a controlled temperature of 600 °C. This neutron irradiation condition is equivalent to a fluence of 5.8×10^{-21} cm^{-2} (E > 1 MeV) and a fast flux of $2-5 \times 10^{-14}$ cm^{-2} s^{-1} (E > 0.1 MeV), corresponding to a displacement damage rate of 3.7×10^{-7} dpa/s. Characterizations of higher dose irradiations (9 dpa) were attempted but analysis was not possible because of the high activity of these neutron irradiation materials and the overwhelming background noise on the position-sensitive single atom detector. These higher doses will require the activity (i.e., the mass) of the specimen to be reduced for future analyses by many orders of magnitude with the use of focused ion beam lift-out methods [7]. This capability was not available for high activity specimens at the time of these experiments.

Atom probe tomography characterizations were performed in voltage pulsing mode with Imago Scientific Instruments LEAP 2017 and LEAP 4000X HR systems for the irradiated and unirradiated conditions, respectively. The LEAP 4000X HR has approximately twice the field of view and is equipped with an energy-compensated reflectron lens to significantly improve the mass resolution. The improved mass resolution enables the Y^{3+} peak on the high mass side of the major iron peak to be fully resolved to the noise floor. Mass peak deconvolution in these nanostructured ferritic alloys are complicated by the presence of molecular ions of TiO, YO and WO, presumably due to the strong affinity between these elements. The nanoclusters were statistically analyzed with the maximum separation method [8] and proximity histograms [9]. A specimen temperature of 50 K, a voltage pulse repetition rate of 200 kHz, and a pulse fraction of 0.2 were used for all atom probe analyses. A maximum separation distance between Ti and O atoms of 0.6 nm was used for the maximum separation method. A size cut off of 10 atoms was used to eliminate clusters due to the random distribution of solute in the ferrite matrix.

RESULTS AND DISCUSSION

Isoconcentration surface representations showing the distributions of the nanoclusters for the unirradiated and neutron irradiated conditions are shown in Figs. 1a and 1b, respectively. The high spatial resolution of the APT technique revealed some non-uniformity in the distribution of the nanoclusters, which complicates comparisons of their number densities. However, these analyses provide a more realistic description of the fine-scale microstructure, and indicate that there are regions of the microstructure where the dislocations will be impeded by the nanoclusters and other regions where the dislocations will move freely. The solute distributions

40

in the unirradiated and neutron irradiated conditions are shown in atom maps in Figs. 2a and 2b, respectively. As demonstrated in previous APT investigations of this and similar nanostructured ferritic alloys, it is evident that the nanoclusters are enriched in titanium, yttrium and oxygen.

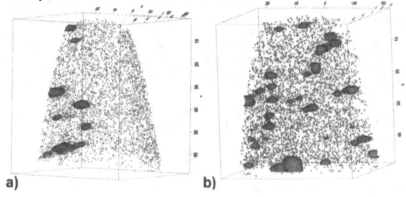

Fig. 1. Ti isoconcentration surfaces of the nanoclusters in a) the unirradiated condition (LEAP 4000X HR) and b) after neutron irradiation to a dose of 3 dpa (LEAP 2017). A small fraction of Ti atoms are also shown to mark the extent of the volume of analysis. The distance scales are in nanometers.

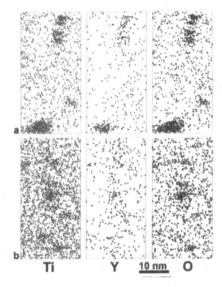

Fig. 2. 5-nm-thick atom maps of the solute distribution in the nanoclusters in a) the unirradiated condition and b) after neutron irradiation to a dose of 3 dpa.

41

The average radii of gyration, l_g, for the Ti and O atoms in the nanoclusters, as estimated by the maximum separation method for the individual and combined solutes, and the number densities of the nanoclusters are given in Table 1 for both the unirradiated and neutron irradiated conditions. The results for l_g(Ti+O), l_g(Ti) and l_g(O) were similar. A comparison of the size distributions of the nanoclusters is shown in Fig 3. No statistical difference in the average size or size distributions of the nanoclusters was found between the unirradiated and irradiated conditions. Due to the large variations in the number densities of the nanoclusters from different regions of the material, only a general comparison is appropriate for this material. To a first approximation, the number density of the nanoclusters was similar in the unirradiated and irradiated conditions.

Table 1. Number density and radius of gyration of the nanoclusters in the 12YWT alloy, as estimated by the maximum separation method for Ti+O atoms

	Unirradiated	Neutron irradiated – 3 dpa
Number density	1.1×10^{24} m^{-3}	9.1×10^{23} m^{-3}
l_g(Ti+O)	0.72 ± 0.26 nm	0.71 ± 0.27 nm
l_g(Ti)	0.72 ± 0.26 nm	0.72 ± 0.27 nm
l_g(O)	0.70 ± 0.26 nm	0.70 ± 0.27 nm

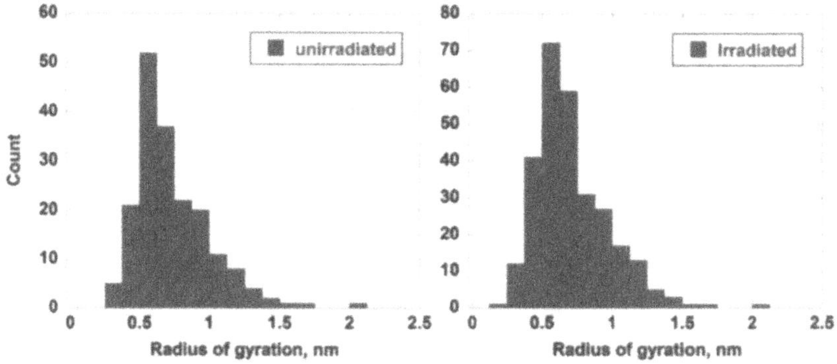

Fig. 3. Comparison of the radius of gyration distributions of the nanoclusters in the unirradiated and 3 dpa neutron irradiated conditions of the 12YWT nanostructured ferritic steel.

For ultrafine features such as these nanoclusters, where the number of atoms in the surface of the nanocluster are the major or a significant proportion of the total number of atoms in the nanocluster, accurate unbiased compositional estimates are primarily limited by locating the precise position of the interface, which is further complicated by the existence of solute profiles for the different elements. Proximity histograms were used to investigate the solute partitioning and distribution within these ultrafine nanoclusters. In this method, the shortest distance of each atom to the nearest isoconcentration surface is determined. The numbers of atoms at these distances are accumulated in a histogram for each element so that the solute concentration as a function of distance from the interface may be estimated. Proximity histograms may be constructed for each nanocluster. However, as the number of atoms associated with each nanocluster is small, the error in each concentration estimate is large. Therefore, the proximity histograms presented here are the averages from all the nanoclusters within the volume of analysis of a specimen, excluding some partially intersected nanoclusters on the exterior surface of the volume. Within the interior of the nanocluster, the number of atoms becomes progressively smaller towards the center of the nanocluster and consequently, the standard error of the measurement increases.

Proximity histograms of the unirradiated and neutron irradiated conditions are shown in Fig. 4. The balance of these concentrations is iron. As observed in previous APT characterizations of these materials, the nanoclusters do not conform to the Y_2O_3 composition of the original yttria powder or any of the coarse oxides (TiO_2, $Y_2Ti_2O_7$, etc.) In addition to the titanium, yttrium and oxygen within the nanoclusters, carbon, nitrogen, and boron were also found to partition to the nanoclusters. These trace elements are present in the material or introduced from adsorbed gases on the surface of the powders, or surface material transferred from the balls and chamber during the ball milling process. Chromium was also observed in the nanocluster and at the nanocluster-matrix interface. This enrichment could be due to segregation to the interface region or rejection of chromium from the interior of the nanocluster. These solutes only account for approximately 40% of the atoms in the nanocluster and the remainder is iron. Neglecting the iron and chromium contributions, the (Ti,Y):(O,C,N,B) ratio was close to 1:1.

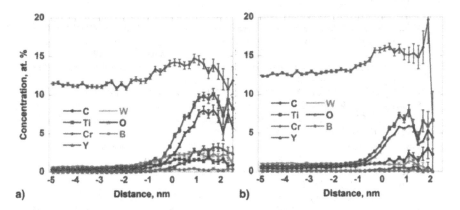

Fig. 4. Proximity histograms of the solute distribution across the nanocluster-matrix interface in a) the unirradiated condition (LEAP 4000X HR) and b) after neutron irradiation to a dose of 3 dpa (LEAP 2017). The nanocluster-matrix interface is located at the origin of the distance axis and nanocluster is at the positive distances. The balance of these concentrations is iron.

A comparison of proximity histograms of the unirradiated and neutron irradiated conditions reveals slightly lower titanium, yttrium, tungsten, and oxygen levels in the nanoclusters in the neutron irradiated material. However, a direct comparison will require data of the neutron irradiated material to be obtained in an energy-compensated instrument due to interference of the tail of the iron peak with the significantly smaller yttrium peak.

CONCLUSIONS

This atom probe tomography investigation of a 12YWT nanostructured ferritic steel has revealed that the nanoclusters are present before and after neutron irradiation to doses of 3 dpa. Neutron irradiation does not change the size or size distribution of the nanoclusters. Although significant region-to-region variations were observed, the average number densities of the nanoclusters were similar in the unirradiated and irradiated conditions. These APT results indicate that the 12YWT nanostructured ferritic steel is remarkably tolerant to relatively high doses of neutron irradiation.

ACKNOWLEDGMENTS

This research was sponsored by the U.S. Department of Energy, Division of Materials Sciences and Engineering. Research at the Oak Ridge National Laboratory SHaRE User Facility was sponsored by the Scientific User Facilities Division, Office of Basic Energy Sciences, U.S. Department of Energy.

REFERENCES

1. M.K. Miller, C.L. Fu, M. Krcmar, D.T. Hoelzer, and C.T. Liu, Shanghai Metallic Material Magazine, 30 (4) 2008 1-6, and Frontiers of Mater. Sci. in China, 3(1), 9 (2009).
2. D. J. Larson, P. J. Maziasz, I. -S. Kim, K. Miyahara, Scr. Mater. 44, 359 (2001).
3. M.K. Miller, E.A. Kenik, K.F. Russell, L. Heatherly, D.T. Hoelzer and P.J. Maziasz, Mater. Sci. Eng. A., A353, 140 (2003).
4. J. H. Schneibel, C.T. Liu, M.K. Miller, M.J. Mills, P. Sarosi, M. Heilmaier and D. Sturm, Scr. Mater., 61, 793 (2009).
5. P. Pareige, M.K. Miller, D.T. Hoelzer, B. Radiguet, E. Cadel and R.E. Stoller, J. Nucl. Mater., 360, 136 (2007).
6. T.R. Allen, J. Gan, J.I. Cole, M.K. Miller, J.T. Busby, S. Ukai, S. Ohtsuka, S. Shutthanandan and S. Thevuthasan, J. Nucl. Mater., 375, 26 (2008).
7. M.K. Miller, K.F. Russell, K. Thompson, R. Alvis and D.J. Larson, Microscopy and Microanalysis, 13(6), 428 (2007).
8. J.M. Hyde and C.A. English, in *Microstructural Processes in Irradiated Materials*, edited by G.E. Lucas, L. Snead, M.A. Kirk, Jr., R.G. Elliman, (Mater. Res. Soc. Symp. Proc. 650, Pittsburgh, PA, 2001) pp. R6.6.1.
9. O.C. Hellman, J.A. Vandenbroucke, J. Rüsing, D. Isheim and D.N. Seidman, Microscopy and Microanalysis, 6, 437 (2000).

Designing Materials for Nuclear Energy I

Mater. Res. Soc. Symp. Proc. Vol. 1215 © 2010 Materials Research Society 1215-V09-04

Densification of Inert Matrix Fuels Using Naturally-occurring Material as a Sintering Additive

S. Miwa[1], M. Osaka, T. Usuki[2] and T. Yano[2]
[1]Oarai Research and Development Center, Japan Atomic Energy Agency, 4002 Narita-cho, Oarai-machi, Higashi-ibaraki-gun, Ibaraki, 311-1393.
[2] Research Laboratory for Nuclear Reactors, Tokyo Institute of Technology, 2-12-1, O-okayama, Meguro-ku, Tokyo 152-8550, Japan

ABSTRACT

We proposed a new concept for densification of inert matrix fuels containing minor actinides. In this concept, magnesium silicates which are both a naturally-occurring material and asbestos waste were used as a sintering additive which protects public health by safely disposing of the asbestos waste. In this study, the effects of magnesium silicate additives on the densification behaviors of MgO, Mo and CeO_2 were experimentally investigated. The densities of MgO and CeO_2 pellets increased with only 1 wt.% additives of $MgSiO_3$ and Mg_2SiO_4. The densities of Mo pellets showed little change with additives.

INTRODUCTION

Inert matrix fuels (IMFs) with a high content of minor actinides (MAs) are currently considered as one promising option for the rapid incineration of MAs in a future fast reactor cycle system [1-5]. IMFs are a composite of MA host phase and an inert matrix (IM). Magnesium oxide (MgO) and molybdenum (Mo) are considered to be promising candidates for an IM material [1-5]. These IMs were selected based on their high manufacturability, good chemical and physical stability, high melting temperature, and relatively high thermal conductivity.

We proposed a new concept for densification of IMFs with Mo and MgO by using asbestos waste as a sintering additive. This concept should contribute especially to the protection of public health (i.e. by disposing of the toxic asbestos waste through its use in the IMFs) in addition to the reduction of environmental burden (i.e. by burning the MAs). Additionally, among the MAs, americium (Am) was found to have high volatility [6], and the lowing the sintering temperature is required to prevent Am loss during sintering. In our concept, magnesium silicates such as enstatite ($MgSiO_3$) and forsterite (Mg_2SiO_4) are used as sintering additive to achieve high-performance IMFs having a high density at relatively low temperature sintering. These substances are naturally-occurring materials and found to be formed by the decomposition of asbestos in a relatively low temperature heat-treatment [7].

In this study, we experimentally investigated the effects of magnesium silicate additives on densification of component materials of IMFs, i.e. host phase and inert matrix, for the purpose of establishing a sophisticated fabrication procedure for MA-containing IMF based on powder metallurgy techniques. Cerium (Ce) oxide was chosen to simulate MAs oxide host phase.

EXPERIMENTAL

The fabrication procedure was based on powder metallurgy techniques. Table 1 shows the fundamental specifications of raw powders. $MgSiO_3$ and Mg_2SiO_4 were milled by planetary ball-milling for 20 min at 600 rpm. MgO (Ube Material Co., Ltd.), Mo (A. L. M. T. Co., Ltd.) or CeO_2 (Sigma-Aldrich Corp.) powders and pre-determined amounts of the magnesium silicates additives were mixed in an agate mortar with a pestle in acetone medium for 15 min. The additive amounts were: MgO powders, 0, 1, 3, 5 and 10 wt.%; Mo or CeO_2 powders, 0, 1, 3, 5 wt.%. Each mixed powder was pressed at 200 MPa into a compact. Sintering tests were carried out at 1673 - 1873 K for 3 h in air atmosphere for MgO or CeO_2 mixed powders, and 4%H_2-Ar atmosphere for Mo mixed powders.

The densification behaviors were characterized by the density and microstructure. The sintered density was obtained from metrological results by using a micrometer and a high accuracy balance. The experimental error of density measurement was plus or minus 0.3 %TD. It was mainly caused by a reading error of scale. The theoretical densities of each material were calculated on the assumption that the sintering additives did not react with matrix material. The microstructure was observed with an optical microscope. The pellets were mounted into holders using epoxy resin, and then were ground and mirror-polished with anhydrous lubricant.

Table 1 Fundamental specifications of raw powders

	MgO	Mo	CeO$_2$	MgSiO$_3$	Mg$_2$SiO$_4$
Purity [%]	99.99	99.9	99.95	-	-
Specific surface area [m^2/g]	12.6	-	-	-	-
Mean particle size [μm]	0.1	2.8-3.7	5	2.5	2.5
Impurity [%]	-	-	-	Al$_2$O$_3$:0.07 CaO: 0.06	Al$_2$O$_3$:2.6 CaO: 0.28

RESULTS AND DISCUSSIONS

Densification of MgO pellets with MgSiO₃ and Mg₂SiO₄

Fig 1 shows the densities of MgO pellets sintered at 1873 K with $MgSiO_3$ ($MgSiO_3$ - MgO) and Mg_2SiO_4 (Mg_2SiO_4 - MgO). The densities of MgO pellets sintered at 1873 K increased with only 1wt.% additive and reached above 95 %TD. The densities of Mg_2SiO_4 - MgO pellets were higher than those of $MgSiO_3$ - MgO pellets. Whereas the densities of Mg_2SiO_4 - MgO pellets slightly increased with increasing the amount of sintering additive, those of $MgSiO_3$ - MgO pellets decreased with increasing the amount of sintering additive. Fig. 2 reproduces ceramographic images of 10 wt.% $MgSiO_3$ - MgO and 10 wt.% Mg_2SiO_4 - MgO pellets. The pore size of 10 wt.% $MgSiO_3$ - MgO was larger than that of Mg_2SiO_4 - MgO. It is believed that the lower density of $MgSiO_3$ - MgO pellets could be attributed to the pore structure.

The densification behaviors for MgO with each magnesium silicate were considered based on the liquid phase sintering mechanisms and/or controlling grain growth effects of MgO matrix by magnesium silicate. Concerning the phase diagram of the Mg-Si-O system, the formations of liquid phases of $MgSiO_3$/Mg_2SiO_4 (33 - 50 at.% SiO_2) and Mg_2SiO_4/MgO (0 - 33 at.% SiO_2) started at about 1823 K and 2133 K, respectively [8]. Therefore, whereas the liquid phase was formed for $MgSiO_3$ - MgO at the sintering temperature, the Mg_2SiO_4 did not melt. For $MgSiO_3$ - MgO, the melting of $MgSiO_3$ would have occurred at 1873 K, and that would result in the formation of liquid phase at the early stage of sintering. Typical sintering behavior in the presence of liquid phase seemed to proceed, and rearrangement of MgO grains via the liquid phase had occurred. The larger amount of sintering additive would lead to the increase of liquid phase. Since it would lead to the increase of densification rate and prevent out-gassing of pores inside the pellet, relatively large pores could be formed and the density would decrease. For Mg_2SiO_4 - MgO, the suppression effects of Mg_2SiO_4 on grain growth of MgO matrix could be predominant, and that could lead to the prevention of the secondary re-crystallization. This controlled grain growth could sweep pores out at the final stage of sintering, and high density could be achieved.

The present results showed that magnesium silicates additives were effective for the densification of MgO. Therefore, it is believed that these additives would make MgO-based IMFs dense using a simple approach with a small amount of magnesium silicates additives and a relatively low sintering temperature.

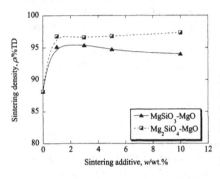

Fig. 1 The densities of MgO pellets sintered at 1873 K as a function of the amount of sintering additive. The densities of MgO pellets increased with only 1wt.% additive. The densities of Mg_2SiO_4 - MgO pellets were higher than those of $MgSiO_3$ - MgO pellets.

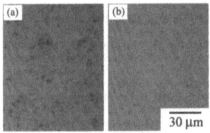

Fig. 2 Ceramographic images of $MgSiO_3$ - MgO (a) and Mg_2SiO_4 - MgO (b) pellets. The pore size of 10 wt.% $MgSiO_3$ - MgO was larger than that of Mg_2SiO_4 - MgO.

Densification of Mo pellets with $MgSiO_3$ and Mg_2SiO_4

Fig 3 shows the densities of Mo pellets sintered at 1873 K with $MgSiO_3$ ($MgSiO_3$ - Mo) and Mg_2SiO_4 (Mg_2SiO_4 - Mo). The sintered densities of Mg_2SiO_4 - Mo pellets sintered at 1873 K showed little change with the additives while the sintered densities of $MgSiO_3$ - Mo pellets were lower.

For Mg_2SiO_4 - Mo, it is believed that Mg_2SiO_4 would show the same effects as for the case of MgO. However, since the sintering temperature of 1873 K was too low for Mo densification, the densification did not progress well. Concerning the formation of liquid phase, in addition to the formation of $MgSiO_3$ liquid phase at 1873 K, there was the possibility that the liquid phase of Mo oxides such as MoO_2 and MoO_3 were formed at low temperature [9]. However, in the sintering atmosphere of 4%H_2 - Ar, since the oxygen partial pressure was very low, the MoO_2 and MoO_3 would not be formed. Therefore, the liquid phase was formed only for $MgSiO_3$ at 1873 K. From the results, it is believed that the effects of $MgSiO_3$ on rearrangement of Mo grains would be small, and densification would not progress well.

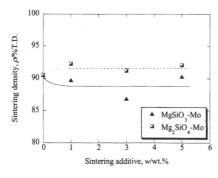

Fig. 3 The densities of Mo pellets sintered at 1873 K as a function of the amount of sintering additive. The sintered densities of Mg_2SiO_4 - Mo pellets showed little change with the additives while the sintered densities of $MgSiO_3$ - Mo pellets were lower.

Densification of CeO_2 pellets with $MgSiO_3$ and Mg_2SiO_4

Fig 4 shows the densities of CeO_2 pellets sintered at 1673 K with $MgSiO_3$ ($MgSiO_3$ - CeO_2) and Mg_2SiO_4 (Mg_2SiO_4 - CeO_2). The densities of CeO_2 pellets sintered at 1673 K increased with only 1 wt.% additives of $MgSiO_3$ and Mg_2SiO_4, and had about 93 %TD. The densities decreased with increasing the amount of additives above 1 wt.% for $MgSiO_3$. Fig 5 reproduces a ceramographic image of 5 wt.% $MgSiO_3$ - CeO_2 pellet. It should be noted that the grain boundary phase, which was a typical structure of liquid phase sintering, was observed at 1673 K.

The reaction between CeO_2 and SiO_2 has been found to take place at 1673 K in air [10] and it can be ascribed to the onset of the reduction of CeO_2 to Ce_2O_3. In addition, it is believed that this temperature could decrease by adding MgO to the Ce-Si-O system surmising from the phase relation of the Mg-Si-O system [8]. Therefore, the liquid phase would form in Ce-Si-Mg-O system below 1673 K and the sintering would proceed. The difference of densification behaviors between $MgSiO_3$ - CeO_2 and Mg_2SiO_4 - CeO_2 would be attributed to the amount of liquid phase. For $MgSiO_3$ - CeO_2, the typical sintering in presence of liquid phase seemed to proceed. However, since the amount of liquid phase increased for 3 and 5 wt.% $MgSiO_3$ so that the pellet shapes were not maintained, the densities which were obtained from metrological results would be low. For Mg_2SiO_4 - CeO_2, it is believed that the formation of liquid phase might be restricted, and the densification could proceed.

From this result, there is a high possibility that the liquid phase would also be formed in Am-Si-Mg-O and Pu-Si-Mg-O systems at a lower temperature based on the similarity of thermochemical properties of CeO_{2-x} with those of PuO_{2-x} and AmO_{2-x} [11]. Therefore, it is believed that these additives would make dense MgO-based IMFs with a relatively low sintering temperature.

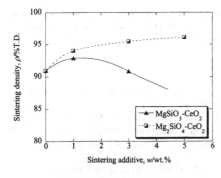

Fig. 4 The densities of CeO_2 pellets sintered at 1673 K as a function of the amount of sintering additive. The densities of CeO_2 pellets increased with only 1 wt.% additives of $MgSiO_3$ and Mg_2SiO_4. The densities decreased with increasing the amount of additives above 1 wt.% for $MgSiO_3$.

20 μm

Fig. 5 Ceramographic image of 5 wt.% $MgSiO_3$ - CeO_2 pellet. The grain boundary phase, which was a typical structure of liquid phase sintering, was observed.

CONCLUSIONS

Fabrication tests of MgO, Mo and CeO_2 with magnesium silicates, i.e. $MgSiO_3$ and Mg_2SiO_4, as a sintering additive were carried out for the purpose of establishing a fabrication method for IMFs having high performance at relatively low sintering temperature.

The density of MgO pellets increased with only 1 wt.% additives of $MgSiO_3$ and Mg_2SiO_4 at 1873 K. The densification mechanism for $MgSiO_3$ and Mg_2SiO_4 would be different at 1873 K. The densification behaviors for $MgSiO_3$ were considered based on the liquid phase sintering mechanisms. Mg_2SiO_4 would have an ability to suppress grain growth of MgO matrix. The densities of Mo pellets showed little change with additives. Although the effects of sintering additives were expected to show the same effects as for MgO, the densification was not achieved with the relatively low sintering temperature. The density of CeO_2 pellets also increased with only 1 wt.% of additives. The liquid phase in the Ce-Mg-Si-O system was formed below 1673 K, and it would enhance the densification.

From the results, there is a possibility that the densification of MgO based IMFs would be enhanced with a small amount of magnesium silicate additives and a relatively low sintering temperature.

ACKNOWLEDGMENTS

This work was supported by Grant-in-Aid for Scientific Research 20560780. We wish to thank Mr. Kosuke Tanaka of JAEA and Mr. Shinichi Sekine of the Nuclear Technology and Engineering Corporation for their invaluable help in these experiments.

REFERENCES

1. M. Osaka, H. Serizawa, M. Kato, K. Nakajima, Y. Tachi, R. Kitamura, S. Miwa, T. Iwai, K. Tanaka, M. Inoue, Y. Arai, J. Nucl. Sci. and Tech. 44, 309 (2007).
2. M. Osaka, S. Miwa, K. Tanaka, I. Sato, T. Hirosawa, H. Obayashi, K. Mondo, Y. Akutsu, Y. Ishi, S. Koyama, H. Yoshimochi, K. Tanaka, in *Energy and Sustainability, Transaction: Ecology and the Environment vol. 105*, edited by C. A. Brebbia and V. Popov (Wessex Institute of Technology, UK, 2007), p. 357.
3. Y. Croixmarie, E. Abonneau, A. Fernandez, R. J. M. Konings, F. Desmouliere, L. Donnet, J. Nucl. Mater. 320, 11 (2003).
4. D. Haas, A. Fernandez, C. Nastren, D. Staicu, J. Somers, W. Maschek and X. Chen, Ene. Conv. Manag. 47, 2724 (2006).
5. M. Osaka, M. Koi, S. Takano, Y. Yamane and T. Misawa, J. Nucl. Sci. Technol. 43, 367 (2006).
6. R. G. Haire, J. Alloys Compd. 213/214, 185 (1994).
7. A. F. Gualtieri and A. Tartaglia, J. Eur. Ceram. Soc. 20, 11(2007).
8. S. Kambayashi and E. Kato, J. Chem. Thermodyn., 15 [8], 701 (1983).
9. L. L. Y. Chang, Trans. Metall. Soc. AIME 230, 1203 (1964).
10. H. A. M. van Hal and H. T. Hintzen, J. Alloys Compd. 179, 77 (1992).
11. C. Sari, E. Zamorari, J. Nucl. Mater. 37, 324 (1970).

Structural Complexity of Nuclear Fuels

Mater. Res. Soc. Symp. Proc. Vol. 1215 © 2010 Materials Research Society   1215-V10-05

Neutron Diffraction Study of the Structural Changes Occurring During the Low Temperature Oxidation of UO_2

Gianguido Baldinozzi[1,2], Lionel Desgranges[3], Gurvan Rousseau[3,1,2]
[1] SPMS MFE, CNRS Ecole Centrale Paris, Chatenay-Malabry, France.
[2] DEN DMN SRMA LA2M MFE, CEA Saclay, Gif-sur-Yvette, France.
[3] DEN DEC SESC LLCC, CEA Cadarache, St. Paul-lez-Durance, France.

ABSTRACT

The oxidation of uranium dioxide has been studied for more than 50 years. It was first studied for fuel fabrication purposes and then later on for safety reasons to design a dry storage facility for spent nuclear fuel that could last several hundred years. Therefore, understanding the changes occurring during the oxidation process is essential, and a sound prediction of the behavior of uranium oxides requires the accurate description of the elementary mechanisms on an atomic scale. Only the models based on elementary mechanisms should provide a reliable extrapolation of laboratory results over timeframes spanning several centuries. The oxidation mechanism of uranium oxides requires understanding the structural parameters of all the phases observed during the process. Uranium dioxide crystal structure undergoes several modifications during the low temperature oxidation that transforms UO_2 into U_3O_8. The symmetries and the structural parameters of UO_2, β-U_4O_9, β-U_3O_7 and U_3O_8 were determined by refining neutron diffraction patterns on pure single-phase samples. Neutron diffraction patterns, collected during the in situ oxidation of powder samples at 483 K were also analyzed performing Rietveld refinements. The lattice parameters and relative ratios of the four pure phases were measured during the progression of the isothermal oxidation. The transformation of UO_2 into U_3O_8 involves a complex modification of the oxygen sublattice and the onset of complex superstructures for U_4O_9 and U_3O_7, associated with regular stacks of complex defects known as cuboctahedra which consist of 13 oxygen interstitial atoms. The structural modifications during the oxidation process are discussed.

INTRODUCTION

The development of advanced nuclear energy systems requires scientific data in order to ensure their safe behaviour throughout the fuel life cycle from fabrication to end of life storage. Neutron diffraction is a very valuable tool for the characterization of UO_2 ceramic nuclear fuel, because neutrons can probe bulk samples and provide reliable information relative to the O sublattice, overcoming the severe limitations of X-ray diffraction experiments. In this in-situ neutron diffraction experiment we have demonstrated the influence of complex oxygen defects [1] and clusters on the oxidation mechanism in UO_2.

The engineering of future advanced nuclear energy systems will require new materials and chemical processing techniques that provide structural integrity, phase stability and process efficiency under extreme conditions of radiation, temperature and corrosive environments for timeframes that in some cases span several thousands of years. Fuel development and qualification spread out over decades, and the process is costly. The basis for fuel development has long been empirical and experience-based because the underlying physical mechanisms are not fully understood. Significantly, there is a substantial lack of knowledge on fundamental material properties even for conventional unirradiated fuel materials, and essentially no information exists for most of the advanced actinide fuels. In fact, many of the required experiments are complex: the materials are not only radioactive, but inherent difficulties exist because the properties of these systems can change sharply with composition, have high vapor pressures, are highly reactive, and may experience internal decay heating. Moreover, reliable fundamental thermochemical and thermophysical information cannot yet be predicted using computational techniques due to the lack of an adequate underlying theory of electronic structure in actinides. The problem of understanding and developing a predictive capability for the evolution of fuels is challenging, even for phenomena that appear simple. In this study, we show how this problem was solved in the case of oxidation of uranium dioxide.

Oxidation is the major risk that has to be taken into account for dry storage of nuclear fuel. During oxidation at temperatures below 600 K, UO_2 undergoes three morphotropic phase transitions, leading to the progressive formation of β-U_4O_9, β-U_3O_7 and finally U_3O_8 [2,3]. A morphotropic transformation can be characterized as the structural change occurring between two adjacent homogeneous phases having different stoichiometries due to the variation in composition. Such transformation generally displays a highly reconstructive mechanism and it is characteristic of the phase diagrams where a series of stoichiometric phases is separated by wide biphasic regions of coexistence where the content of the neighboring phases varies continuously with concentration. This approach differs essentially from the concentration-wave approach used for describing the homogeneous structures of solid solutions. Only the crystal structures of UO_2, U_4O_9 and U_3O_8 were already determined. We performed high-resolution in situ neutron diffraction experiments [4] to determine the changes in the UO_2 structure during oxidation; more specifically, the structure of β-U_3O_7 was determined and refined.

EXPERIMENT

The in situ isothermal oxidation experiment (483 K) was performed on the D2B high resolution neutron powder diffractometer at ILL. The sample was put into a cylindrical vanadium sample holder and dry air was forced to flow through it to ensure a stable oxidizing atmosphere during the experiment. Since a dense sample would prevent achieving the optimum conditions for the uniform oxidation of the sample, UO_2 powder was dispersed in the cylindrical sample holder using fused silica fibers to ensure a dry airflow within the sample. UO_2 and U_4O_9 phases cannot be easily distinguished in our experiment because of a severe overlap of the fluorite basic peaks. A U_3O_7 single-phase is observed after about 10 hours. U_3O_8 formation occurs after about 17 hours.

The crystal structures of U_4O_9 (actually $U_{256}O_{572}$) and U_3O_7 (actually $U_{256}O_{592}$) are still closely related to the fluorite structure of UO_2. The neutron diffraction patterns display superstructure reflections witnessing an ordering of the extra O atoms. Since the pioneering works of Bevan, a large number of different models were proposed for these structures. Cooper and Willis [5] recently solved the crystal structure of β-U_4O_9.

In this structure, the oxygen incorporation into UO_2 fluorite structure results in the formation oxygen interstitial clusters named cuboctahedra. In the large U_4O_9 cell (64 times larger than the original UO_2 one) there are 12 isolated cuboctahedra.

We have recently solved the crystal structure of U_3O_7. The M_6X_{37} cluster is virtually identical with the basic M_6X_{32} unit of fluorite, so that the former atomic cluster can be incorporated readily within a fluorite matrix. In tveitite, this occurs in an ordered manner, In spite of the denser packing of the anions in the filled cuboctahedron, there are no unusually short interatomic distances (see Fig. 1).

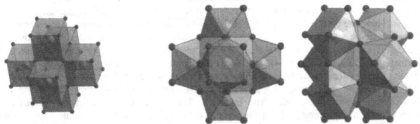

Fig. 1. Left Panel: the atomic arrangement in the ideal fluorite structure consisting of 6 MX_8 cubes sharing their edges. Right panel: two representations of the M_6X_{37} group, the new structural element, produced from the previous one by a simple rotation mechanism that involves four more cations; it consists of six MX_8, square antiprisms sharing their corners to enclose a cuboctahedron where an extra X cation sits at the centre. The atomic group M_6X_{37} was evidenced by Bevan [6,7] in minerals having fluorite related structures as tveitite $Ca_{14}Y_5F_{43}$.

To solve the structure of U_3O_7 we have built three models in the fourfold fluorite cubic space group, centring the 16 cuboctahedra at particular positions on the threefold axis, leading to resulting structures similar to the ones already proposed for U_4O_9 by Bevan [8]. Then we have described the atomic positions in these initial structures using the tetragonal subgroup of highest symmetry compatible with the observations (I-42d). The three models were refined and the model where the cuboctahedra are centred at x=1/16 in the prototype cubic structure gives the best results. It is important to point out that the actual structure of U_3O_7 involves a significant tilt and deformation of the cuboctahedron, almost respecting a -4 local symmetry for the atomic cluster [4].

Fig. 2. Left Panel: Crystal structure of the β phase of U_4O_9 at 400K. Middle panel: Crystal structure of U_3O_7 at 523K. Right panel: crystal structure of U_3O_8 at 533 K.

The formation of U_3O_7 is associated with the increase of the number and the spatial redistribution of these cuboctahedra in the crystal structure (Fig. 2, panel a and b). The main difference between U_4O_9 and U_3O_7 is that the insertion of 16 cuboctahedra in the large cell cannot be done without sharing some oxygen atoms between the deformed square antiprisms surrounding each cuboctahedron. These contacts involve significant distortions of the square antiprisms. U_3O_7 marks the hyperstoichiometric limit above which the fluorite-like phases in the UO_{2+x} system become unstable and the cuboctahedra are no longer able to efficiently arrange the excess oxygen atoms. Beyond this limit the formation of U_3O_8 occurs (Fig. 2, panel c).

There is a clear difference between the crystal lattice and the symmetry at this transition. Experiments on single crystals [9] suggest a topotactic growth of U_3O_8 onto U_3O_7, providing strong hints that the main structural changes at the interface still affect the oxygen sublattice. The formation of U_3O_8 possibly takes place when the topological frustration imposed on the (111) dense planes by the cuboctahedra is no longer sufficient to prevent the rearrangement from the ABC stacking of U_3O_7 into the AA stacking of U_3O_8. From a phenomenological viewpoint, this is very similar to a martensitic phase transition. The formation of U_3O_8 seems to start at a threshold level depending both on the oxygen concentration and on the initial stress in the sample.

DISCUSSION AND CONCLUSION

These structural refinements provide accurate models for the structures of uranium oxides from UO_2 to U_3O_8 and they enable to follow the structural modifications occurring during the oxidation at the atomic level. These structures provide a benchmark for the refinements of the diffraction patterns obtained during the isothermal oxidation where several phase coexist in the sample.

Although having different space groups, these uranium oxides share several similarities. This is especially true when we look at the structure as a stacking of (111) layers. This point will be illustrated, first by recalling the crystalline relationships between UO_2 and U_3O_8 proposed by Allen et al. [10], in which these dense planes are significant.

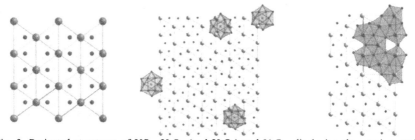

Fig. 3. Projected structures of UO_2, U_4O_9 (and U_3O_7) and U_3O_8, displaying the uranium and oxygen positions in the (111)$_c$ dense planes. U atoms are grey, O atoms are red. In the central panel, the O atoms occupying fluorite positions are red, those belonging to the cuboctahedron are blue.

UO_2 structure is based upon the close-packing of ions into hexagonal layers normal to the main diagonal of the fluorite cell. Each layer contains only one kind of atom: either U or O. These hexagonal layers are stacked in an ABC fashion, so that the fourth layers lies immediately over the first and so forth. The regular ABC stacking of hexagonal layers consisting of alternate elements gives rise to the rocksalt structure. In UO_2, a fluorite structure, half of the U layers are missing.

In spite of the different chemical formulae, the hexagonal U sublattice in U_3O_8 is very similar to the one in the UO_2 fluorite structure and a topotactic growth is then possible. The hexagonal (indeed, the average structure is hexagonal but it presents a local orthorhombic distortion) (001) planes of U_3O_8 oxide can efficiently grow on the (111) layers of UO_2. Nevertheless, the structural matching between (111) planes in the fluorite structure and the (001) planes in U_3O_8 is limited to the U sublattice. Moreover, the stacking sequence of the (001) planes in U_3O_8 consists of only a single kind of layers (A) that repeats. The main differences between the two structures arise from the anion arrangement. While in UO_2 all the O atoms lie in separate layers, in U_3O_8 most of the O atoms now sit in the same U layer creating a pentagonal pattern around each U atom. A limited amount of O atoms still sits in the interlayer on top of each U atom creating pillars that connect these layers. These U-O-U bonds are shorter than the U-O bonds in UO_2 (2.077 instead of 2.368Å) and they form the height of pentagonal bipyramids. Whether the U-O bonds in the interlayer are short, the distance between two U layers is now larger than in UO_2 (4.152 instead of 3.157Å) and it is responsible for most of the volume increase observed during the phase transformation from UO_2 to U_3O_8. On the other hand, U-U distances in the layer are not much longer than those in UO_2 though O atoms are now sharing this same layer.

The description of U_4O_9 and U_3O_7 structures can shed some light on the structural changes affecting the anion sublattice occurring during the oxidation. The incorporation of oxygen cuboctahedra in the UO_2 structure starts the deep modification of the anion sublattice. The main structural feature of U_4O_9 structure is the existence of 12 cuboctahedra, each one made of 13 O atoms: 6 neighbor U atoms, sharing the normal cation sublattice of the fluorite structure, have a square antiprism coordination instead of the normal cubic one. The interstitial O atoms describing the cuboctahedron faces deplete the anion sites in the fluorite structure and they introduce new O atom layers closer to the U layers. On the other hand, the atoms at the centre of the cuboctahedra lie midway between two U layers, in a position that it is possible to identify with the O atoms forming the pillars connecting the layers in U_3O_8 structure. When a thick slab formed by a U layer and the new O atom layers consisting of interstitial O atoms are projected along <111>, this projected slab is already very similar to the pentagonal pattern existing in U_3O_8 structure (Figure 3, right panel). Therefore, the formation of cuboctahedra starts to produce some of the structural features responsible for the larger structural modifications occurring in U_3O_8.

When U_3O_7 is formed, 16 cuboctahedra, noticeably tilted and deformed are incorporated in the fluorite structure. It is important to point out that none of the 16 cuboctahedra in U_3O_7 is at the same position of one of the 12 cuboctahedra in U_4O_9. This fact witnesses the great mobility of O atoms in these structures. Therefore, the O sublattice efficiently reorganizes via the creation and destruction of cuboctahedra. In U_4O_9, none of the atoms of the square antiprisms surrounding the cuboctahedra is shared. But the 16 cuboctahedra in U_3O_7 cannot be efficiently distributed in the unit cell avoiding the mutual sharing of some of the O atoms defining the square antiprisms. These contacts produce short U-O distances (< 2.1 Å) responsible for the large increase of the bond valence sum (BVS) [11] values of the neighbouring U atoms compared to the corresponding values in the U_4O_9 structure. These short U-O bonds induced both inside or outside the cuboctahedra are mainly associated with the tetragonal distortion of the U_4O_9 lattice and to the cuboctahedra tilt. In this process, each cuboctahedron loses its threefold axis and it becomes more elongated along the c_t axis.

It is important to stress that the metastable nature of U_3O_7 implies that, provided enough time is given to reach the thermodynamic equilibrium at the annealing temperature, it will transform in a mixture of U_4O_9 and U_3O_8. Therefore, it is not possible to exclude the possibility that the cuboctahedron configurations determined in this experiment may be dependent on kinetic parameters. At present, no study was performed to check whether the structural features of U_3O_7 depend on these parameters. The accumulation of local deformations in U_3O_7, due to the frustrated environments of the square antiprisms, may be incompatible with the long-term existence of the cuboctahedra and, above a given threshold it may trigger its transformation into U_3O_8.

REFERENCES

1 G. Petot-Ervas, G. Baldinozzi, P. Ruello, L. Desgranges, G. Chirlesan and C. Petot. Mat. Res. Soc. Symp. Proceedings **824** 217–222 (2004)

2 R.J. McEachern and P. Taylor, J. Nucl. Mat. **254**, 87-121 (1998)

3 G. Rousseau, L. Desgranges, J.-C. Nièpce, J.-F. Bérar, and G. Baldinozzi, Mat. Res. Soc. Symp. Proceedings **802** 3–8 (2003)

4 L.Desgranges, G. Baldinozzi, G. Rousseau, J.C. Nièpce, G. Calvarin, Inorg. Chem., **48** 7585-7592 (2009).

5 R. I. Cooper and B. T. M. Willis, Acta Crystallogr. A **60**, 322-325 (2004).

6 D. J. M. Bevan, J. Strahle and O. Greis, J. of Solid State Chemistry **44**, 75-81 (1982).

7 D. J. M. Bevan, J. Strahle and O. Greis, Acta Crystallogr. A **36** 889-890 (1980).

8 D. J. M. Bevan, I.E. Grey and B.T.M. Willis, J. of Solid State Chemistry **61**, 1-7 (1986).

9 L. Quémard, L. Desgranges, V. Bouineau, M. Pijolat, G. Baldinozzi, N. Millot, J.C. Nièpce, A. Poulesquen, J. of the European Ceramic Soc. **29** 2791–2798 (2009)

10 G. C. Allen, P. A. Tempest, Proceedings of the Royal Society of London. Series A, Mathematical and Physical Sciences **406**, 325-344 (1986)

11 I.D. Brown and D. Altermatt, Acta Crystallogr. B41 (1985) 244-247

Mater. Res. Soc. Symp. Proc. Vol. 1215 © 2010 Materials Research Society   1215-V10-06

Microstructure and Thermophysical Characterization of Mixed Oxide Fuels

Franz J. Freibert, Tarik A. Saleh, Fred G. Hampel, Daniel S. Schwartz, Jeremy N. Mitchell, Charles C. Davis, Angelique D. Neuman, Stephen P. Willson, and John T. Dunwoody

Los Alamos National Laboratory, Los Alamos, NM 87545, U.S.A.

ABSTRACT

Pre-irradiated thermodynamic and microstructural properties of nuclear fuels form the necessary set of data against which to gauge fuel performance and irradiation damage evolution. This paper summarizes recent efforts in mixed-oxide and minor actinide-bearing mixed-oxide ceramic fuels fabrication and characterization at Los Alamos National Laboratory. Ceramic fuels $(U_{1-x-y-z}Pu_xAm_yNp_z)O_2$ fabricated in the compositional ranges of $0.19 \leq x \leq 0.3$ Pu, $0 \leq y \leq 0.05$ Am, and $0 \leq z \leq 0.03$ Np exhibited a uniform crystalline face-centered cubic phase with an average grain size of 14µm; however, electron microprobe analysis revealed segregation of NpO_2 in minor actinide-bearing fuels. Immersion density and porosity analysis demonstrated an average density of 92.4% theoretical for mixed-oxide fuels and an average density of 89.5 % theoretical density for minor actinide-bearing mixed-oxide fuels. Examined fuels exhibited mean thermal expansion value of $12.56 \times 10^{-6}/°C$ for temperature range $(100°C<T<1500°C)$ and ambient temperature Young's modulus and Poisson's ratio of 169 GPa and of 0.327, respectively. Internal dissipation as determined from mechanical resonances of these ceramic fuels has shown promise as a tool to gauge microstructural integrity and to interrogate fundamental properties.

INTRODUCTION

An understanding of the thermophysical properties of oxide fuels, in conjunction with knowledge of their composition and microstructure, is critical for improving materials processing methods, understanding fuel performance, populating databases from which models may be developed, and validating modeling and simulation. Mixed-oxide ceramic fuels of composition $(U_{1-x}Pu_x)O_2$, where $0.1 \leq x \leq 0.3$ Pu, occupy a region of the (U, Pu, O) ternary phase diagram which is single phase cubic fluorite structure at ambient temperatures and above [1] which can accommodate significant O vacancies and structural defects. This paper documents the fabrication of mixed-oxide (MOX) and minor actinide-bearing mixed-oxide (MAMOX) ceramic fuels; details the microstructural characterization of these ceramics including thermal etching, image analysis for grain and porosity studies, and electron backscatter diffraction; and describes the results of thermophysical properties determinations including lattice parameters, density, thermal expansion, and elastic moduli. This combination of datasets - microstructure, composition, phase inclusions, and thermophysical properties - have been correlated with processing variations in Pu-Ga alloys and successfully utilized in modeling the thermodynamic properties of actinide alloys [2]. These experimental studies strengthen the understanding of actinide material structures and properties while in non-equilibrium conditions.

EXPERIMENT

Various lots of MOX and MAMOX fuels were available for this work. Ceramic fuels $(U_{1-x-y-z}Pu_xAm_yNp_z)O_2$ were fabricated in the compositional ranges of $0.19 \leq x \leq 0.3$ Pu, $0 \leq y \leq 0.05$ Am, and $0 \leq z \leq 0.03$ Np. All compositions began with the combining of constituent oxides with 1 weight % polyethylene glycol binder as a binder. These mixtures were then milled in a SPEX™ mill using a zirconium oxide jar and ball for 15 minutes. The milled powders were loaded into a 6.0 mm diameter die and pressed at 120MPa for 10 seconds. Subsequently, binder burned out was accomplished during a 4 hour 450°C thermal treatment under flowing argon. The pellets were then sintered at temperatures ranging from 1625°C to 1750°C for 4 hours. Based on published studies of sintering in an atmosphere of moist argon-hydrogen gas mixture [3], the nominal stoichiometry of the oxide had a predetermined oxygen-to-metal ratio (O/M) of 2.00. The quality of the as-fabricated 5mm x 5mm right circular cylinder pellets was determined by visual inspection, an examination to ensure minimal edge chipping, fracturing, surface voids, etc. The MOX and MAMOX fuels characterization results presented here include those for the following compositions: optical microscopy for $(U_{0.80}Pu_{0.20})O_2$ and $(U_{0.76}Pu_{0.19}Am_{0.03}Np_{0.02})O_2$, porosity and x-ray diffraction of $(U_{0.80}Pu_{0.20})O_2$ and $(U_{0.76}Pu_{0.19}Am_{0.03}Np_{0.02})O_2$, elemental mapping of $(U_{0.76}Pu_{0.19}Am_{0.03}Np_{0.02})O_2$, elastic moduli determined for $(U_{0.80}Pu_{0.20})O_2$, thermal expansion measured for $(U_{0.76}Pu_{0.19}Am_{0.03}Np_{0.02})O_2$, and density for a compositional series of MOX and MAMOX samples. This work comprises a subset of MOX and MAMOX fuels characterization activities which have been and continue to be conducted at Los Alamos National Laboratory.

Ceramographic preparation and analysis was conducted on as-fabricated pellets which were initially mounted in a sacrificial 3.0 mm thick epoxy chip as an aid to handling. These specimens were then ground starting with a 240 soft grit and transitioned to a 600 soft grit. Polishing was performed on a vibratory polisher through a sequence of steps from 6 μm diamond, then 1 μm diamond, and finally 0.05 μm colloidal alumina slurry. The epoxy was carefully removed and the specimen was rinsed in ethanol. The ethanol left a residue of fine, adherent round residual debris spots of diameter ≤ 5 μm. Thermal etching was implemented on both MOX and MAMOX specimens to reveal grain structure sufficient for grain size and porosity measurements. The specimen were heated to 1375°C and held for 15 minutes in a regeneration gas (Ar + 6% H_2). The process resulted in pronounced voids and excellent grain boundary visibility. Specimens were then examined using optical microscopy. Grain morphology was quantified by manually tracing the grain outlines and using Image Pro Plus software to process the trace image.

Porosity (or void volume fraction) was quantified by taking advantage of the differing grey value of the pores in comparison to the grains as the pores are revealed when voids are incised during sample serial sectioning. By selecting appropriate grey level bands, images were created that contained only pores which were then analyzed for size and shape. The presence of residual stains from the ethanol rinse was a complication for this specimen set. Image Pro Plus software filtering tools adjusted for morphological parameters such as size and roundness of the stains allowed them to be easily filtered out of the analysis.

Optical microscopy of a $(U_{0.76}Pu_{0.19}Am_{0.03}Np_{0.02})O_2$ sample revealed a region (~20 μm in diameter) of anomalously small grains (~2 μm) surrounded by grains of otherwise typical size (~14 μm). In response, a qualitative x-ray mapping of uranium, plutonium, americium, and neptunium via electron microprobe analysis was conducted, in order to determine if segregation of one of the constituent actinides had occurred. (Elemental mapping will not reveal oxygen, but based on actinide ceramic chemistry there is no expectation segregation in the metallic or other compound form.) The sample was carbon coated for charge compensation in preparation for electron microprobe analysis with a JEOL 8200. Elemental maps were background subtracted and collected for each element at a 256 x 256 mm area with a spatial resolution of 0.16 mm, an accelerating voltage of 6 KeV, and a 150 nA probe current. Red-green-blue maps were also generated to facilitate analysis of element location and collocation.

Lattice parameters of MOX and MAMOX specimen were measured using a Thermo ARL Xtra X-ray powder diffractometer. Copper K_{α} radiation was used for the x-ray analysis. Specimens were milled by mortar and pestle to minimize grain orientation effects during data collection. To eliminate the spread of contamination, the specimens were wrapped in plastic film which produces broad polymeric peaks in the 15-27° two-θ range. NIST SRM 660 LaB_6 powder was mixed in with the ground specimens as an internal lattice parameter standard.

The immersion density technique was employed by which specimen density is derived from the buoyant force acting on the specimen immersed in a liquid of known density. This technique has been modified to include the corrections necessary for self-heating samples [4]. A Mettler Toledo balance Model AT201 electronic semi micro-balance was used to obtain dry and wet weights. The immersion liquid used in this study was the 3M Corporation product Fluorinert™ FC-43. The density of the immersion fluid is determined before each series of measurements using a NIST lead silica glass standard. As a control charting and error estimating activity, a 1cm³ control sample of Invar is utilized to ensure consistent performance of the bath over time and temperature. The temperature of the immersion fluid was monitored with a platinum resistor having good sensitivity (0.3 ohm/K) at 290K. Utilizing a four probe resistance measurement, sensitivity on the level of +/- 0.005 K is easily achieved. This temperature data is used to correct for any changes in wet sample mass that occurs from the deposition of heat into the immersion fluid when measuring self-heating materials. Samples were lightly cleaned prior to being measured to remove any loose oxides on the surface. Using care in all material handling steps, the accuracy of the reported density determination is typically ± 0.001 g/cm³ for samples over 10 g and ± 0.01 g/cm³ for samples less than 1g. The mass of these samples varied from approximately 0.9g to 2.0g.

Resonant ultrasound spectroscopy (RUS), a method of determining the elastic moduli of materials from the mechanical resonant mode frequencies of a sample [5], was employed in this study. This technique works well with the cylindrical pressed fuel pellet (aspect ratio ~1) where codes are available to predict resonant mode frequencies and has been successfully employed in other actinides elemental and alloy research [2, 6] where implementation in a glovebox environment is necessary. Knowing the dimensions and weight of a sample, its density, and rough elastic moduli for the material, allows one to predict its resonant mode frequencies. Minimization in the difference between the numerical values for the experimentally measured mode frequencies and predicted mode frequencies, as the estimated elastic moduli are varied, results in a very accurate measurement of the elastic moduli.

The thermal expansion behavior of fuels specimens were determined using dilatometry. The measurement was performed using a Netzsch 402C pushrod dilatometer with a high-temperature graphite furnace. The thermocouple was protected using an alumina cover, and the sample was isolated from the pushrod and tube using alumina disks. The furnace was flushed for 30 minutes using ultra-high purity Ar and a flow rate of 30 ml/min was used throughout the measurement. All measurements were performed at heating rates of 2°C/minute. To calibrate the dilatometer, a Netzsch sapphire standard was measured under the same conditions to evaluate accuracy of the instrument at these temperatures. As with RUS, this technique has been successfully employed for actinides composite materials, phase transformations, and structural stability studies [2, 7] in a glovebox environment.

DISCUSSION

The optical photomicrograph image in figure 1 of $(U_{0.76}Pu_{0.19}Am_{0.03}Np_{0.02})O_2$ is typical of the ceramic fuels specimens being studied. Grain boundaries are quite evident. Brighter round features are the adherent rounded residual debris spots of $\leq 5\mu m$ diameter deposited when the ethanol rinse evaporated. This fact was used to differentiate between residue and voids during the porosity analysis. Pores were not included in these measurements. Grains were traced in such a way that their boundaries passed through pores on double boundaries and met approximately in the center of pores at triple and higher junctions. This has the effect of slightly increasing the apparent grain size, but gives a grain size corresponding to a fully dense specimen, which is a natural reference. Boundaries of the photomicrographs were kept clear, i.e. grains that overlapped boundaries were not included in the analysis. The grain morphology parameters collected were mean diameter, major and minor axes of best fit ellipse, aspect ratio. At 500X, the images typically contained 200 – 300 grains for analysis. The results of the grain size analysis show the expected log-normal distribution in histogram plots. The average diameter was 14 ± 8 µm, which is comparable to published values for similarly prepared MOX fuels [3]. This value is useful for comparison to the pre-sintered powder size to understand how the precursor powders change size and shape as the material consolidates. The grains were somewhat elongated, with an average aspect ratio of 1.5.

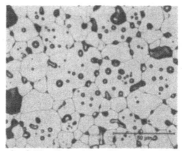

Figure 1. Micrograph of $(U_{0.76}Pu_{0.19}Am_{0.03}Np_{0.02})O_2$ composition ceramic fuel.

The principal goal of the porosity measurements was to determine the void volume fraction. In making the assumption that the pores are isotropically distributed in the specimens and have shape which does not deviate extremely from equiaxed, the pore aerial fraction which is measured in 2-dimensional micrographs is equivalent to the true 3-dimension volume fraction of voids. As such, the void volume fraction derived from 4 micrograph datasets for $(U_{0.80}Pu_{0.20})O_2$ and $(U_{0.76}Pu_{0.19}Am_{0.03}Np_{0.02})O_2$ were 0.10±0.01 and 0.17±0.01, respectively. The voids which were observed were non-communicating and homogeneously distributed.

64

For the specimen of $(U_{0.76}Pu_{0.19}Am_{0.03}Np_{0.02})O_2$ which exhibited an anomalously fine-grained region, grain tracing was used to quantify the grain diameters in this region. In general, the grains in this region exhibited an average grain size of 2.1 ± 0.8 μm and showed no unusual morphologies. Elemental maps revealed that this region was comprised of primarily neptunium - assumed to be in the dioxide form or NpO_2. Based on color mapping, some of the grains included other actinides; however, overall the small grained region is mostly neptunium.

The use of an internal standard coupled with multi-parameter fitting routines available in Materials Data Inc. Jade software allowed for a highly accurate lattice parameter determination. These oxides have the fluorite crystal structure. The $(U_{0.80}Pu_{0.20})O_2$ composition exhibited lattice parameter a_0 of 5.4575 Å and the $(U_{0.76}Pu_{0.19}Am_{0.03}Np_{0.02})O_2$ composition exhibited lattice parameter a_0 of 5.4599 Å. In comparison, the published MOX lattice parameter for $(U_{0.81}Pu_{0.19})O_{2.00}$ which is 5.456Å.[1]

The theoretical density ρ of MOX as a function of PuO_2 composition follows the literature derived relationship ρ=10.95+0.55c where c is the volume fraction PuO_2 [1, 8]. For MAMOX, the theoretical values for the density were estimated by one-for-one atomic substitution of Np for U and Am for Pu. Based on data collected by immersion density, the $(U_{0.80}Pu_{0.20})O_2$ showed an average density 10.404 ± 0.018g/cm^3 or 93.56% theoretical density, $(U_{0.70}Pu_{0.30})O_2$ showed an average density 10.095 ± 0.078g/cm^3 or 91.15% theoretical density, $(U_{0.65}Pu_{0.30}Am_{0.03}Np_{0.02})O_2$ showed an average density 10.079 ± 0.018g/cm^3 or 89.45% theoretical density, and $(U_{0.75}Pu_{0.20}Am_{0.03}Np_{0.02})O_2$ showed an average density 9.824 ± 0.474g/cm^3 or 88.58% theoretical density. Samples of compositions $(U_{0.80}Pu_{0.20})O_2$ and $(U_{0.76}Pu_{0.19}Am_{0.03}Np_{0.02})O_2$ which also received microstructural examination exhibited densities of 91.77% and 85.55%, respectively. In comparing those data with the void volume fraction numbers of 0.10 ± 0.01 and 0.17 ± 0.01 from porosity measurements, there is very good agreement between these techniques. For all MOX and MAMOX compositions, the higher densities correlated with the lower sintering temperature of 1625°C as compared with 1750°C.

Results from immersion density and porosity analysis where used to inform the elastic moduli analysis. For comparison, literature values of Young's modulus E as a function of porosity for UO_2 exhibits a E=223-500v behavior [9] and $(U_{0.8}Pu_{0.2})O_2$ exhibits a E=210-420v behavior [10], where v is the volume void fraction. Measured immersion density and these elastic moduli values were used as starting points for the elastic moduli determination. Based on a measured void fraction porosity of 0.08 for these materials, these samples were observed to have Young's modulus of 169 GPa and Poisson's ratio of 0.327 and an average fitting error of 1.32%. These data compare well with literature values for $(U_{0.8}Pu_{0.2})O_2$ with 0.08 void fraction having Young's modulus of 176 GPa and Poisson's ratio of 0.311. However, using these values as starting points revealed that the initial resonance peak failed to be exhibited on all samples, thus leading to fitting errors higher than usual. Noteworthy is the observation that some of the samples exhibited little or no resonance behavior typically seen in cracked materials, while others resonated but were missing key resonance peaks, which made fitting the data challenging. Such mechanical behavior coupled with results from microstructural and morphological analyses has supported studies of heterogeneity and phase mixtures composites [2].

Because the thermal expansion measurements were conducted in an inert atmosphere, there was no control of the O stochiometry and therefore limited certainty that the O/M remained at 2. Thermal expansion measurements on samples of $(U_{0.76}Pu_{0.19}Am_{0.03}Np_{0.02})O_2$ displayed several slope changes as a function of temperature, notably at about 850 °C and 1394 °C. These inflection points may have been an indication that the single phase ceramic may have partially decomposed under heating at temperatures greater than 800°C. However, between these inflections, the measured thermal expansion coefficients compare with those determined by Martin [11] for $(U_{1-x}Pu_x)O_2$, where $0.2 \leq x \leq 0.3$ Pu as follows: $\bar{\alpha}_{Exp}=10.91x10^{-6}$ °C^{-1} compared with the literature value $\bar{\alpha}_{Std}=10.51x10^{-6}$ °C^{-1} for 100 °C $\leq T \leq$ 800°C and $\bar{\alpha}_{Exp}=16.14x10^{-6}$ °C^{-1} compared with the literature value $\bar{\alpha}_{Std}=11.94x10^{-6}$ °C^{-1} for 800°C $\leq T \leq$ 1300 °C. Over the entire range of data collection (100 °C $\leq T \leq$ 1500 °C), the measured thermal expansion coefficients was $\bar{\alpha}_{Exp}=12.56x10^{-6}$ °C^{-1} as compared with the literature value $\bar{\alpha}_{Std}=11.11x10^{-6}$ °C^{-1}.

CONCLUSIONS

An understanding causal relationship of processing and chemical composition to microstructure and thermophysical properties is crucial to process optimization and performance predictability. Now that the challenges of conducting these measurements are better understood, a comprehensive suite of tools to measure and to evaluate MOX and MAMOX fuel materials can be applied with confidence. These data and an improved capability for microstructural characterization will greatly inform future MOX and MAMOX ceramic fuels fabrication and processing. Bulk elastic properties and thermal expansion characteristics support an understanding of the equation of state of these materials and provide a performance indicator under extreme conditions.

Other outcomes from this work include the recognition that the accurate determination of high temperature thermodynamic properties of these ceramics, such as thermal expansion, heat capacity, and elastic moduli will require the development of a protocol for monitoring and controlling O/M changes at temperature. Also, measuring the internal mechanical energy dissipation as observed in mechanical resonance peaks shows promise as a tool to classify and quantify defects impacting sample quality and performance, such as microcracking, voids, and compositional heterogeneity in MOX and MAMOX ceramic fuels.

ACKNOWLEDGMENTS

This research was supported by the Nuclear Energy Fuel Cycle Research and Development Program of the U.S. Department of Energy.

REFERENCES

1. T.L. Markin, and R.S. Street, *J. Inorg. Nuc. Chem.* **29**, 2265 (1967).
2. F.J. Freibert, J.N. Mitchell, T.A. Saleh, D.S. Schwartz, "Thermophysical Properties of Coexistent Phases in Plutonium"; accepted for publication in the *IOP Conf. Series: Materials Science and Engineering (2009)*; Los Alamos National Laboratory Publication. No. LAUR-09-00291 (2009).
3. R. Güldner and H. Schmidt, *J. Nucl. Mater.* **178**, 152 (1991).
4. F.J. Freibert, D. Dooley, D. Miller, Los Alamos National Laboratory Publication No. LAUR-05-9007, 2005.
5. A. Migliori, J.P. Baiardo, T.W. Darling, and F.J. Freibert in *Experimental Methods in the Physical Sciences: Modern Acoustical Techniques for the Measurement of Mechanical Properties* **39**, edited by M. Levy, H.E. Bass, and R. Stern (Academic Press, San Diego, 2001) p. 189.

6. A. Migliori, C. Pantea, H. Ledbetter, I. Stroe, J.B. Betts, J.N. Mitchell, M. Ramos, F.J. Freibert, D. Dooley, S. Harrington, C.H. Mielke, *JASA* **122**, 1994 (2007).
7. J.N. Mitchell, F.J. Freibert, D.S. Schwartz, and M.E. Bange, *J. Nuc. Mat.* **385**, 95 (2009).
8. Theoretical density data derived from lattice constants published by R.E. Skavdahl and T.D. Chikalla in *Plutonium Handbook: A Guide to the Technology*, Vol I and II, edited by O.J. Wick (American Nuclear Society, La Grange Park, Illinois, 1980), p. 261.
9. A. Padel and Ch. de Novion, *J. Nucl. Mater.* **33**, 40 (1969).
10. A.W. Nutt, A.W. Allen, and J.H. Handwerk, *J. Am. Ceram. Soc.* **53**, 205 (1970).
11. D.G. Martin, *J. Nuc. Mat.* **152**, 94 (1988).

Designing Materials for Nuclear Energy II

Mater. Res. Soc. Symp. Proc. Vol. 1215 © 2010 Materials Research Society   1215-V11-01

Radiation Tolerance and Disorder – Can They Be Linked?

Karl R Whittle, Mark G Blackford, Katherine L Smith, Gregory R Lumpkin [1], and Nestor J Zaluzec [2]

1 - Institute of Materials Engineering, Australian Nuclear Science and Technology Organisation, PMB1, Menai, NSW 2234, Australia
2 - Electron Microscopy Center, Argonne National Laboratory, Argonne, IL

ABSTRACT

The future expansion of nuclear power provides materials challenges that are not easily overcome, for example the safe immobilisation of nuclear waste is an important component in any future expansion of nuclear power. The use of ceramic-based materials, as opposed to borosilicate glasses, is now being investigated widely. For an oxide to have uses as a waste form the effects of radiation damage must be understood. As part of a long-term research programme the effects on radiation tolerance of a range of ordered and disordered materials are discussed. The results suggest that materials with a high tolerance for disorder, whether it be chemical or structural will tolerate radiation damage effects better than those that are highly ordered.

INTRODUCTION

How materials behave when used as materials for the safe storage of active nuclear waste undergoing radioactive decay has been studied as a means to improve current, and develop future nuclear materials[1]. It is not always possible to undertake experiments using radioactive isotopes, the timescales can be too long to be viable, in many cases it is often quicker to use ion-beam irradiation to model the damage processes[2-4]. While such a technique may not always give 'real world' results it does allow 'model' systems to be studied systematically, where results can be used to understand currently relevant materials, and provide a means by which simulations of damage can be assessed. One area of discussion in the literature is the effect of structure versus composition, and in particular the effects of chemical/structural disorder within a system on the resistance to, or toleration of, radiation damage. As part of this ongoing process we have studied a range of systems with varying degrees of disorder:

i) $La_{2-x}Y_x(Hf/Zr)_2O_7$ – a pyrochlore to fluorite transition with increasing Y content, studied by neutron and electron diffraction prior to irradiation.[5, 6]
ii) $La_{2/3}TiO_3$ and $La_2Ti_2O_7$ – perovskite based materials. [7, 8]
iii) La_2TiO_5 – an orthorhombic structure with atypical Ti and La co-ordinations (5 and 7 respectively)
iv) Y_2TiO_5, Yb_2TiO_5 and $YbYTiO_5$ – disordered pyrochlore materials
v) $Y_2Ti_{2-x}Sn_xO_7$ – ordered pyrochlores with random Ti/Sn mixing on the B-site.[9, 10]

The systems were chosen so that where possible minimal change was made in either composition or structure, thus allowing the effects of disorder to be correlated with the change in radiation damage tolerance.

Both $La_{2-x}Y_x(Hf/Zr)_2O_7$ and $Y_2Ti_{2-x}Sn_xO_7$ adopt the standard pyrochlore structure, shown in Figure 1a, but with differing degrees of disorder with the change in compositions. The $La_{2-x}Y_x(Hf/Zr)_2O_7$ series was studied by neutron and electron diffraction and found to disorder with increasing Y, ultimately adopting a defect fluorite structure – although with evidence of long range ordering incommensurate with unit-cell size[5]. The second pyrochlore system $Y_2Ti_{2-x}Sn_xO_7$ was found by ^{89}Y MAS NMR to show no evidence for mixing across the A and B sites, and with no ordering of Ti or Sn on the B-site [9, 10].

The systems Y_2TiO_5, Yb_2TiO_5 and $YbYTiO_5$ adopt a structure identical in nature to pyrochlore, Figure 1a, but with inherent cationic disorder on the B-site. When the composition is reformulated to match that found in pyrochlore i.e. $Y_2Ti_{1.33}Y_{0.67}O_{6.67}$, Y is found on both the A and B sites, and deficiency within the O lattice, whose order is as yet undetermined.[11]

The La_2TiO_5 system adopts a structure, shown in Figure 1b, with atypical co-ordinations for both the La (7 as a corner missing cube) and Ti (5 trigonal bipyramid), whereas $La_{2/3}TiO_3$ and $La_2Ti_2O_7$ are related to perovskite [7, 8]. As can be seen the system studied show evidence of both highly ordered and disordered structures.

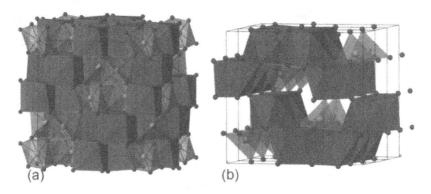

Figure 1 – (a) the standard pyrochlore structure showing the octahedrally coordinated B-site (hollow octahedron) and the scalenohedron A-site (filled in polyhedron). (b) structure adopted by La_2TiO_5 with the 5 coordinated Ti shown in the hollow polyehdron and the 7 coordinated La polyhedra shown filled in.

EXPERIMENTAL

The samples were prepared predominantly by the calcination of oxides at high temperatures, ~ 1500°C, until phase pure. The exact nature of synthesis can be found in the references[2, 5-9]. Once prepared the samples were fully characterized using SEM, TEM, XRD and neutron diffraction analysis where appropriate.

Electron Microscopy and Microanalysis

Samples were checked for purity by scanning electron microscopy and microanalysis (SEM-EDX) using a JEOL 6400 operating at 15 kV. Microanalyses were obtained using a Noran Voyager energy dispersive spectrometer attached to this microscope. The instrument was operated in standardless mode; however, the sensitivity factors were calibrated for semi-quantitative analysis using a range of synthetic and natural standard materials. Spectra were usually acquired for 500 seconds and reduced to weight percent oxides using a digital top hat filter to suppress the background, a library of reference spectra for multiple least squares peak fitting and full matrix corrections.

Selected samples were additionally analysed using transmission electron microscopy and microanalysis (TEM-EDX). TEM samples were prepared by crushing small fragments in methanol and collecting the suspension on holey carbon coated copper grids. Samples were using a JEOL 2000FXII TEM operated at 200 kV and calibrated for selected area diffraction over a range of objective lens currents using a gold film standard. The compositions of the grains were checked by EDX analysis using a Link ISIS energy dispersive spectrometer attached to the TEM. The k-factors required for the quantitative thin film analyses were determined from a range of synthetic and natural standard materials. Spectra were usually acquired for 600 seconds and processed using a digital top hat filter to suppress the background, a library of reference spectra for multiple least squares peak fitting, and a Cliff-Lorimer ratio procedure to reduce the data to weight percent oxides (details are given in Lumpkin et al[12]).

Neutron Diffraction

Neutron diffraction was used to obtain information on both cations and anion location in these materials as is more accurate in the determination of oxygen parameters in the presence of heavier metal atoms than X-ray diffraction. Constant wavelength neutron diffraction patterns were collected, from densely packed powder in thin-walled cylindrical vanadium cans, using the C2-DualSpec diffractometer at the Chalk River National Laboratory, Canada, operating at an incident wavelength of 1.3283Å (determined using an Al_2O_3 standard). Diffraction patterns were collected over the angular range $5°$-$120°$ 2θ, with an angular resolution of $0.1°$, corresponding to a d-spacing range of $0.767 - 15.226$ Å, and at a temperature of 298K. Further information is provided in Whittle et al[5].

X-ray Diffraction

Samples were examined using a Bruker D8 diffractometer with monochromatic Cu Ka1 radiation (1.5406 Å) and a tuned Sol-X detector. The angular range used was 10-$80°$ 2θ, corespoding to a d-spacing range of 1.198-8.835 Å, with a step size of $0.01°$, and counting time of 5 s.

In Situ Ion Beam Irradiation

The samples were irradiated, as crushed grains dispersed on a holey-carbon coated Cu grids, using the IVEM-TANDEM facility at the Argonne National Laboratory, with 1 MeV Kr^{2+} ions, in situ within a Hitachi 9000-NAR transmission electron microscope. A Gatan liquid

73

helium (LHe) cooling stage and a Gatan heating stage were used to conduct experiments at temperatures between 30 and 275 K. and between room temperature and 650 K respectively. The temperature monitored was that of the holder close to the sample. Due to the conductive nature of both the carbon support film, and Cu grid the estimated temperature rise by the sample is low.

To prevent irradiation induced temperature changes the irradiations were carried out using a fluence of 6.25 x 10^{11} ions cm^{-2} s^{-1}, with the sample angled midway between the incident ion and electron beams, 15° to the vertical. During irradiation the electron beam was switched off to prevent synergistic effects of electron and ion beams interacting within the sample.

Multiple grains were irradiated simultaneously, with the point of amorphisation being the point at which no Bragg intensities were visible in the selected area diffraction pattern. This is referred to as the critical fluence (Fc) for the sample at the temperature of measurement.

The data were analysed to determine the critical temperature (Tc) the temperature above the rate of crystalline recovery is equal to the damage rate. This temperature is determined by least squares numerical analysis of the recorded data. The equation used for this process is shown below (1), and is used to determine Tc, energy of activation for crystalline recovery (Ea), and the critical fluence at 0 K (Fc_0) which is proportional to the damage cross section for the system under analysis.

$$Fc = \frac{Fc_0}{1 - \exp\left[\left(\frac{Ea}{k_b}\right)\left(\frac{1}{Tc} - \frac{1}{T}\right)\right]} \tag{1}$$

Where k_b is Boltzmann's constant. It has been well reported in the literature that values for E_a given by this equation are often under estimated, when compared with defect migration energies calculated for oxides. As has been suggested previously by Weber these low values are reflective of the process of amorphisation being controlled by more than kinetic processes. A second method for determining E_a is therefore used based on amorphisation cross-sections, ion-flux and jump frequency, equation 2.

$$Ea = Tc \left[k_b \ln\left(\frac{Fc_0 \nu}{\phi}\right) \right] \tag{2}$$

Where T_c and Fc_0 are those values determined from equation 1, ν is the effective jump frequency in these samples, an estimated value of 10^{12} was used, and ϕ is the ion flux (6.25 x 10^{11} ions cm^{-2} s^{-1}). However, for this process to be valid good estimates of Fc_0, T_c (both can be determined using equation 1 or making estimates from the recorded data) and known ion flux (ϕ) are required.

RESULTS

74

La$_{2-x}$Y$_x$(Hf/Zr)$_2$O$_7$

Three samples in this system were irradiated, La$_2$Zr$_2$O$_7$, La$_2$Hf$_2$O$_7$ and La$_{1.6}$Y$_{0.4}$Hf$_2$O$_7$. The full series for x = 0 to 2 in both systems, were characterized using neutron and electron diffraction, and showed that La$_2$Zr$_2$O$_7$, La$_2$Hf$_2$O$_7$ and La$_{1.6}$Y$_{0.4}$Hf$_2$O$_7$ were single-phase pyrochlore compositions. It was impossible to study the cationic disorder using neutron and electron diffraction alone, for an in-depth explanation please see Whittle et al and Wuensch et al. The diffraction work additionally showed that Y$_2$Zr$_2$O7 and Y$_2$Hf$_2$O$_7$, expected to be a disordered defect fluorite, showed evidence for longer range ordering, but with distances incommensurate with the lattice parameter.

The critical temperature obtained for La$_2$Zr$_2$O$_7$ was 339(49) K, while that for the similar system La$_2$Hf$_2$O$_7$ was significantly higher 563(10) K. The large change ~225 K, suggests that Hf based materials recover more slowly than the Zr equivalent, as can be seen by an increase in critical temperature for La$_2$Hf$_2$O$_7$ compared with La$_2$Zr$_2$O$_7$, shown in Figure 2. However, the replacement by 0.4 formula units of La (fpu) with Y decreases the critical temperature back down to 474(20) K, shown in Figure 2. As there was no evidence in these materials for ordering across the La/Y site, the addition of Y effectively increases the disorder in the material. It can therefore be expected that increasing the Y content further would increase the tolerance of the materials. Unpublished results we have obtained agree, as it proved difficult to amorphise La$_{1.2}$Y$_{0.8}$Hf$_2$O$_7$. Previous studies of Y$_2$Zr$_2$O$_7$ have shown the material to be highly resistant to damage under these conditions.

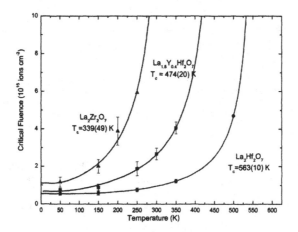

Figure 2 – Plot showing the recorded data and fitted curve obtained for La$_2$Zr$_2$O$_7$, La$_2$Hf$_2$O$_7$ and La$_{1.6}$Y$_{0.4}$Hf$_2$O$_7$, damaged with 1 MeV Kr^{2+} ions. The change from La$_2$Zr$_2$O$_7$ to La$_2$Hf$_2$O$_7$ increases the critical temperature for amorphisation.

La₂/₃TiO₃, La₂Ti₂O₇ and La₂TiO₅

La$_{2/3}$TiO$_3$ and La$_2$Ti$_2$O$_7$ are either perovskite or perovskite-related in structure, with a large degree of site disorder. In the La$_{2/3}$TiO$_3$ system, to ensure charge balance there are vacancies randomly distributed on the A-site (La). The La$_2$Ti$_2$O$_7$ however, is a structure with 4 La sites, and 4 Ti sites, shown in Figure 3a. They are arranged in a manner that gives rise to very asymmetric TiO$_6$ octahedra, and a range of La-O coordination polyhedra. The final structure La$_2$TiO$_5$ is highly ordered with atypical coordination for both La (7) and Ti (5 – trigonal bipyramid). The recorded critical temperatures for La$_{2/3}$TiO$_3$ and La$_2$Ti$_2$O$_7$ are 852(5) and 840(9) K respectively, whereas La$_2$TiO$_5$ was found to be 1027(9) K, shown in Figure 3b. If the disordered and ordered nature of the structures is compared, i.e. La$_2$TiO$_5$ is highly ordered, whereas La$_{2/3}$TiO$_3$ is disordered, a trend can be seen. In the two systems with the lower Tc, the material, once damaged, has many pathways by which it can recover. For example in La$_{2/3}$TiO$_3$ there are many formally vacant positions by which the La can locate during recovery. In La$_2$TiO$_5$ there is extra vacant interstitial space in the lattice, by which an atom can reside during recovery but ultimately it must adopt a crystallographic site appropriate to the species, i.e. La returns to a La site, and Ti to a Ti site. Therefore, for this to occur rapidly, i.e. the critical temperature, a higher temperature thus higher energy of activation for recrysallisation, for rapid reformation will be required, than for those systems where there are more positions that can be adopted.

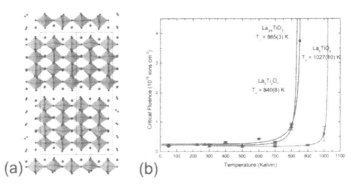

(a)   (b)

Figure 3 – (a) representation of the La$_2$Ti$_2$O$_7$ unit cell showing the block nature of the structure, the polyhedra are TiO$_6$, the isolated atoms are the LaO$_x$ polyhedra (isolated dots). (b) the critical fluence as a function of temperature curves obtained from La$_2$Ti$_2$O$_7$, La$_{2/3}$TiO$_3$ and La$_2$TiO$_5$. The critical temperatures (Tc) were obtained by applying equation (1).

Y₂TiO₅, Yb₂TiO₅ and YbYTiO₅

As outlined previously these systems adopt structure practically identical with pyrochlore, the major difference being the inherent disorder on the B-site. For example Y$_2$TiO$_5$ equates to Y$_2$(Ti$_{1.33}$Y$_{0.67}$)O$_{6.67}$. This disorder changes the obtained critical temperature for these systems. Collected results for Y$_2$Ti$_2$O$_7$ has shown the Tc to be 660(33) K, whereas the obtained Tc's for the others are Y$_2$TiO$_5$ ~589(18) K, Yb$_2$TiO$_5$ ~456(27) K, and YbYTiO$_5$ ~ 466(2) K, shown in Figure 4. As these systems are inherently disordered, there are more pathways available for recovery from damage. This corresponds with $Y_2Ti_2O_7$ that has fewer pathways for damage and does not inherently disorder across the two-cation sites. [9, 10] These systems therefore show that as disorder increases the ability to recover from damage also increases, thus agreeing with the previous observations above.

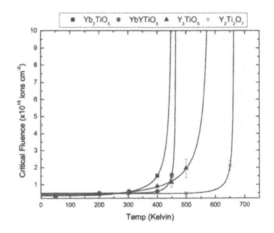

Figure 4 – Plot showing the recorded data and fitted curve obtained for Yb_2TiO_5, $YbYTiO_5$, Y_2TiO_5 and $Y_2Ti_2O_7$ (sample prepared and irradiated under identical conditions), damaged with 1 MeV Kr^{2+} ions.

$Y_2Ti_{2-x}Sn_xO_7$

This series of samples all adopt the pyrochlore structure, and with no evidence of mixing between the Y and Ti/Sn sites. The Ti and Sn are randomly distributed on the B-site with no evidence for any ordering [9, 10]. As the amount of Sn increase it would be expected the amount of disorder increases to a maximum at Y_2TiSnO_7, and decrease on either side. Using the observations above as a guide it would be expected that the Tc would be at a minimum close to the middle of the compositional range, with an increase on both sides. However, this does not occur, the obtained Tc's for this range of samples consistently decreases with increasing Sn content, until a point is reached by which the sample does not amorphise and remains crystalline, shown in Figure 5.

As can be seen with the samples above an increase in disorder, whether it be site or chemical can often lead to a reduction in the critical temperature for amorphisation that in turn implies the material is more resistant, or tolerant of damage. For example those systems where there is only 'ideal' atomic arrangement, e.g. $Y_2Ti_2O_7$ the Tc is higher, whereas in those system where there many arrangements, e.g. Y_2TiO_5, the Tc is lower. However, an increase in disorder cannot account for all of the decrease in Tc, other factors play a role. For example in $Y_2Ti_{2-x}Sn_xO_7$ the systems with the most expected disorder, and thus lowest Tc's would be $Y_2Ti_{1.2}Sn_{0.8}O_7$ and $Y_2Ti_{0.8}Sn_{1.2}O_7$. In the data analysed here the drops from $Y_2Ti_2O_7$ through to

77

$Y_2Ti_{0.8}Sn_{1.2}O_7$, at which point the system becomes highly resistant to damage and does not amorphise under these conditions. One explanation for the stability of $Y_2Sn_2O_7$ and $Y_2Sn_{1.6}Ti_{0.4}O_7$ is due to the high covalent nature of the bond [13], which increases the overall stability of the system compared with $Y_2Ti_2O_7$. Therefore an increase in disorder is not the only driving force for increasing the stability of systems to radiation damage.

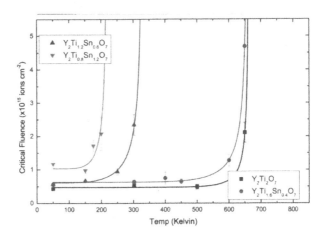

Figure 5 - Plot showing the recorded data and fitted curve obtained for $Y_2Ti_{2-x}Sn_xO_7$ damaged with 1 MeV Kr^{2+} ions. $Y_2Ti_{0.4}Sn_{1.6}O_7$ could not be amorphised, and $Y_2Sn_2O_7$ had previously been reported to not amorphise under these conditions.

CONCLUSIONS

Disordered systems, whether it be chemical or structural disorder, tend to be more tolerant of radiation damage, and recover quickly back to a crystalline state. Those systems with a high degree of order, e.g. only one structure can be ideally adopted, will tend to have a higher Tc as the recovery process requires a much higher energy of activation. However, other factors continue to affect the stability of materials to radiation damage, and the internal nature of the bond cannot be discounted in any prediction of radiation tolerance.

ACKNOWLEDGMENTS
The authors thank the IVEM-Tandem Facility staff at Argonne National Laboratory for assistance during the ion irradiation work. The IVEM-Tandem Facility is supported as a User Facility by the U.S. DOE, Basic Energy Sciences, under contract W-31-10-ENG-38, and the help of Ed Ryan and Pete Baldo in running this facility. We acknowledge financial support from the Access to Major Research Facilities Programme (a component of the International Science Linkages Programme established under the Australian Government's innovation statement, Backing Australia's Ability). Part of this work was supported by the Cambridge-MIT Institute

(CMI), British Nuclear Fuels Limited (BNFL), and EPSRC Grant (EP/C510259/1) to G.R. Lumpkin.

REFERENCES

1.   W. J. Weber, R. Ewing, C. R. A. Catlow, T. Diaz de la Rubia, L. W. Hobbs, C. Kinoshita, H. Matzke, A. T. Motta, M. Nastasi, E. K. H. Salje, E. R. Vance and S. J. Zinkle, *Journal of Materials Research*, 13, 1434-1484 (1998)
2.   G. R. Lumpkin, K. L. Smith, M. G. Blackford, K. R. Whittle, E. J. Harvey, S. A. T. Redfern and N. J. Zaluzec, *Chemistry of Materials*, 21, 2746-2754 (2009)
3.   G. R. Lumpkin, K. L. Smith, M. G. Blackford, B. Thomas, K. R. Whittle, N. Marks and N. J. Zaluzec, *Physical Review B*, 77, 214201 (2008)
4.   R. Ewing, W. J. Weber and J. Lian, *Journal of Applied Physics*, 95, 5949-5971 (2004)
5.   K. R. Whittle, L. Cranswick, S. Redfern, I. P. Swainson and G. R. Lumpkin, *Journal of Solid State Chemistry*, 182, 442-450 (2009)
6.   G. R. Lumpkin, K. R. Whittle, S. Rios, K. L. Smith and N. J. Zaluzec, *Journal of Physics: Condensed Matter*, 16, 8557-8570 (2004)
7.   Z. Zhang, G. R. Lumpkin, C. Howard, K. Knight, K. R. Whittle and K. Osaka, *Journal of Solid State Chemistry*, 180, 1083-1092 (2007)
8.   E. Harvey, K. R. Whittle, G. R. Lumpkin, R. Smith and S. Redfern, *Journal of Solid State Chemistry*, 178, 800-810 (2005)
9.   S. W. Reader, M. R. Mitchell, K. E. Johnston, C. J. Pickard, K. R. Whittle and S. E. Ashbrook, *Journal of Physical Chemistry C*, 113, 18874-18883 (2009)
10.   S. Ashbrook, K. R. Whittle, G. R. Lumpkin and I. Farnan, *Journal of Physical Chemistry B*, 110, 10358-10364 (2006)
11.   M. Petrova, A. Novikova and R. Grebenshchikov, *Inorganic Materials*, 39, 509-513 (2003)
12.   G. R. Lumpkin, K. L. Smith, M. G. Blackford, R. Gieré and C. T. Williams, *Micron*, 25, 581-587 (1994)
13.   B. Kennedy, B. Hunter and C. Howard, *Journal of Solid State Chemistry*, 130, 58-65 (1997)

Mater. Res. Soc. Symp. Proc. Vol. 1215 © 2010 Materials Research Society 1215-V11-04

Production of Layered Double Hydroxides for Anion Capture and Storage

Jonathan D. Phillips and Luc J. Vandeperre
Department of Materials, Imperial College London, Prince Consort Road, London, SW7 2AZ, United Kingdom.

ABSTRACT

Technetium has a long half life of up to 2.13×10^5 years. It is separated from liquid waste streams with tetraphenylphosphonium bromide [1], which upon degradation releases Tc as the pertechnetate anion, $TcO4^-$. Pertechnetate is highly mobile in groundwater and it is therefore highly desirable to capture and immobilise this anion within a solid for interim and ultimately long term storage. Layered Double Hydroxide (LDH) materials are known to possess excellent anion sorption capabilities due to their structure which consists of ordered positively charged sheets intercalated with interchangeable hydrated anions. The composition can be tailored to produce suitable precursors for ceramic phases by varying the divalent and trivalent cations and the anions. LDHs with the general formula $Ca_{1-x} (Fe_{1-y}, Al_y)_x (OH)_2 (NO_3)_x . nH_2O$ were produced by a co-precipitation method from a solution of mixed nitrates. Calcination leads to the formation of Brownmillerite $Ca_2(Al,Fe)_2O_5$ like compounds for temperatures as low as 400°C, this is close to the lowest temperature at which Tc is known to volatilise (310.6 °C Tc_2O_7). It was shown that after calcining up to 600°C, the LDH structure is recovered in water allowing rapid ion capture to occur. This suggests these materials have potential for both capture and as a storage medium for Tc.

INTRODUCTION

In the United Kingdom, one of the waste streams produced during the reprocessing of spent nuclear fuel at the Sellafield site is a medium active liquor rich in radionuclides including technetium. To convert this liquor to a solid the radionuclides are precipitated by the addition of sodium hydroxide, a flocculating agent and tetraphenylphosphonium bromide (TPPB) [2]. The resultant floc is separated from solution by ultrafiltration and incorporated into a cementitious waste form, before being stored in a repository. The addition of TPPB allows technetium to be removed from the waste stream with an efficiency of 97%.[1]

However it is known that in alkaline environments, such as those found in the cement, TPPB can degrade by alkaline hydrolysis resulting in the release of the pertechnetate anion [2]. The pertechnetate anion does not bind well with soils and as such is highly mobile in groundwater [3]. Technetium-99 is of significant environmental concern because it is a weak beta emitter produced in significant quantity (6% of all fission products) and has a long half life of 2.13×10^5 years.

Layered double hydroxides (LDHs) are family of materials renowned for their anion exchange and capture properties [4]. Furthermore they can be synthesized with a broad range of compositions which could make them ideal as precursors for other phases.

LDHs are structurally similar to the minerals Brucite, $Mg(OH)_2$, and Portlandite, $Ca(OH)_2$, in which a central divalent metal cation is surrounded by six hydroxyl groups in an octahedral configuration. Through edge sharing the octahedra form into large two dimensional sheets with hydroxyl terminated surfaces. In LDHs, a fraction of the divalent cations is replaced by trivalent cations, which results in the formation of a net positive charge on the sheets. This charge is neutralized by the incorporation of hydrated anions in the interlayer galleries. For calcium containing LDHs, known as hydrocalumite-like materials, the large ionic radius of the calcium cation enables it to bond with a water molecule in the interlayer and thus become seven coordinated [5].

As with normal clays, the anions in the interlayer can exchange with other anions present in the solution the LDH is in contact with. It is possible to enhance this effect by first driving off the interlayer anions by calcining the material at moderate temperatures. During heating the layered structure is lost but it reforms when the calcined material is exposed to water (memory effect). As the layered structure reforms, new charge balancing anions are incorporated and this leads to very efficient anion capture [4]. Calcining at higher temperatures leads to the loss of the memory effect, and conversion to a stable ceramic oxide.

The aim of this work is to produce a waste form for capturing pertechnetate from solution, which can subsequently be thermally converted into a stable ceramic phase with technetium incorporated within its crystal structure. LDHs containing Ca, Al and Fe, which upon thermal conversion should yield Brownmillerite-like, $Ca_2(Al,Fe)_2O_5$, compounds, are used as a model system and the work reported here describes the production and characterization of the LDH materials as well as their response to heating.

EXPERIMENT

Production of LDHs

Layered double hydroxides containing Ca, Al and Fe were produced by co-precipitation. A 2M solution of the nitrate salts of Ca, Fe and Al ($Ca(NO_3)_2.4H_2O$, $Fe(NO_3)_3.9H_2O$ and $Al(NO_3)_3.9H_2O$ was produced by dissolving the salts in the desired stoichiometric ratio in 15ml of deionised water. Dropwise addition of this solution, with vigorous stirring to a vessel containing 25ml of 0.85M $NaNO_3$ solution resulted in the precipitation of an LDH phase. The pH of the $NaNO_3$ solution was maintained at a pH greater than 12 by the simultaneous addition of a 3M NaOH solution. To improve the crystallinity of the samples hydrothermal aging treatments were employed in the mother liquor either at room temperature for 24-48 hours or at elevated temperature for shorter periods of time. The precipitate was retrieved from the solution by vacuum filtration (0.65μm, Millipore) and rinsed carefully with deionised water to flush unwanted salts from the product. The filtrate was subsequently dried at either room temperature or at 80°C.

Characterization

The chemical composition of the solids formed during precipitation was determined using inductively coupled plasma optical emission spectroscopy (ICP-OES, Perkin Elmer) of solutions obtained by nitric acid digestion of the precipitates. A Scanning Electron Microscope (JEOL, JSM5610LV) was employed to examine the morphology of the powders. X-ray diffraction was carried out using a Philips PW1720 powder x-ray diffractometer with Cu K_α radiation. Initially, XRD sample preparation consisted of depositing a film of powder onto a substrate by allowing a dispersion in acetone to dry on the substrate. However, acetone uptake by the materials was found to alter the interlayer spacing, and unless indicated otherwise results shown were obtained by loading the powder in a recessed sample holder.

Investigation of the effect of heating

The changes that occur upon heating the material were characterized by thermal analysis and X-ray diffraction. Thermal analysis was carried out using a combined TGA/DSC (Netzsch, STA 449 C). The samples were heated at a rate of 10 °C/minute from room temperature to 950°C in an open platinum crucible under a flowing air atmosphere. X-ray diffraction was also used to investigate whether the LDH structure reforms when the calcined material is re-exposed to water. Hereto, typically 0.1g of calcined material was added to 5 ml demineralised water and retrieved again by filtration after 18 hours.

RESULTS AND DISCUSSION

X-ray diffraction revealed that an LDH type phase forms: the indexed pattern in Figure 1 is similar to patterns reported for other calcium containing LDHs, see e.g. [6]. The pattern is dominated by intense peaks at low angles, due to preferential orientation of the samples during analysis. The morphology of the LDH crystallites is shown in Figure 2. It is clear that large sheet like structures have formed, as expected for this family of materials.

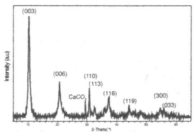

Figure 1 XRD pattern obtained for $Ca_{0.67}Al_{0.165}Fe_{0.165}(OH)_2.(NO_3)_{0.33}.nH_2O$. In addition to the LDH peaks, there is also a peak at 29.4° 2θ, which is characteristic of $CaCO_3$.

Figure 2 Secondary electron image of the hexagonal platelets and flaky morphology observed for a sample of $Ca_{0.7}Al_{0.15}Fe_{0.15}(OH)_2.(NO_3)_{0.3}.nH_2O$.

Compositional analysis using ICP-OES indicated the final product closely matched the nominal starting compositions. For a sample with nominal compostion x=0.33 and Fe:Al ratio of 1 the

ICP results were x=0.36 and 1.07. The discrepancy in x may be due to a machine limitation for detecting calcium.

When hydrothermal aging was carried out at 50°C for up to 20 hours the crystallinity of the LDH improved. For periods greater than 20 hours, $Ca_3Al_{1.54}Fe_{0.46}[(OH)_4]_3$ and $Ca(OH)_2$ were identified in addition to the LDH phase. The intensity of the $Ca_3Al_{1.54}Fe_{0.46}[(OH)_4]_3$ signal increased with aging time (20-140 hours) as shown in Figure 3. Hence the product formed should not be used above 50°C for prolonged times as the loss of the layered structure would also lead to the loss of the captured anions.

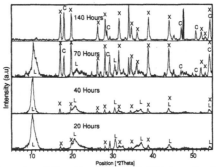

Figure 3 XRD patterns of samples of $Ca_{0.7}Al_{0.15}Fe_{0.15}(OH)_2(NO_3)$ aged at 50°C for different periods of time, from 20-140 hours. Where L indicates $Ca_{0.7}Al_{0.15}Fe_{0.15}(OH)_2(NO_3)$ LDH, X is $Ca_3Al_{1.54}Fe_{0.46}[(OH)_4]_3$ and C is $Ca(OH)_2$.

Effect of calcination

Layered double hydroxides normally exhibit multi-stage weight losses, as the different components volatilize. The thermal analysis curves (TGA and DSC) obtained for a sample of CaAlFe-LDH with x=0.3 are presented in Figure 4 and are in good agreement with literature [7].

Four major weight loss events are observed in the TGA, the first occurring below 200°C is attributed to the loss of interlayer water. At 200-300°C simultaneous dehydroxylation of the sheets and conversion of the nitrate present to nitrite occurs, NO_3-NO_2. Subsequently at 500°C the nitrite is driven out of the sample. Carbonate ions present in the starting solution may have lead to the formation of calcite which subsequently decomposes at approximately 650°C. Based on the weight loss, the extent of calcite contamination is limited to ~12 wt%, which bodes well for industrial scale up or use as an extremely simple process yields a good quality product.

The XRD patterns obtained for samples calcined at a variety of temperatures (not shown) demonstrated that Brownmillerite and calcium oxide were formed at temperatures as low as 400 °C. Below this temperature some of the original LDH structure was retained. While 400 °C is still above the boiling point of Tc_2O_7 (310 °C [8]), it is much lower than the processing temperature for alternative ceramics that have been proposed for incorporation of Tc [8] which therefore increases the likelihood that Tc can be incorporated into the ceramic structure.

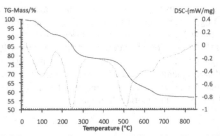

Figure 4 TGA(–) and DSC($\cdots$) curve for a sample of $Ca_{0.7}Al_{0.15}Fe_{0.15}(OH)_2.(NO_3)_{0.3}.nH_2O$

Reconstruction procedure

The XRD patterns obtained for a sample before calcination, after calcination and after rehydration are presented in Figure 5. During heating the layered structure is lost and Brownmillerite forms, Figure 5b. Upon rehydration, however, the layered structure reforms for calcinations up to 600 °C. The rehydrated sample has an (003) peak at a higher angle than in the untreated sample. It is thought that this is due to the capture of the carbonate or hydroxyl anion during reformation of the LDH structure as no other ions were present. For higher temperature calcinations up to 950°C the memory effect was only partially retained as additional phases such as $Ca_3Al_{1.54}Fe_{0.46}[(OH)_4]_3$ were formed during rehydration of the sample.

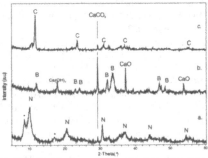

Figure 5 XRD patterns obtained for a) nitrate intercalated CaAlFe LDH (N), * effect of acetone exposure during XRD sample preperation, b) sample of 'a' heat treated to 600°C B=Brownmillerite, CaO=Calcium Oxide, c) sample of 'b' rehydrated in H_2O C=carbonate or hydroxyl intercalated CaAlFe LDH.

CONCLUSIONS

Layered double hydroxides containing Ca, Fe and Al were produced by a coprecipitation method. Hydrothermal aging can be employed to increase the crystallinity of the material, however if exposure to high aging temperature (50°C) is prolonged beyond 20 hours the LDH structure transforms slowly into $Ca_3Al_{1.54}Fe_{0.46}[(OH)_4]_3$. Thermal analysis revealed that despite the absence of any special precautions to prevent it, carbonate contamination was limited to 12wt% calcite. Therefore large scale production and use of this material will not require complicated procedures. Thermal conversion of these phases to Brownmillerite like phases $(Ca_2(Al,Fe)O_5)$ and calcium oxide was possible at temperatures as low as 400°C. Such a low conversion temperature might enable Tc to dissolve into the ceramic host rather than volatilising. It has been demonstrated that the memory effect exists for these compositions from 400°C up to 600°C. Up to 950°C the memory effect occurs to a limited extent as $Ca_3Al_{1.54}Fe_{0.46}[(OH)_4]_3$ is formed in addition to the LDH phase during rehydration.

ACKNOWLEDGMENTS

Funding by UK's Engineering and Physical Sciences Research Council through the DIAMOND consortium is gratefully acknowledged.

REFERENCES

[1]   J. Reed, "Technetium to go," *Nuclear Engineering International,* vol. 49, pp. 14-17, 2004.

[2]   S. Aldridge, P. Warwick, N. Evans, and S. Vines, "Degradation of tetraphenylphosphonium bromide at high pH and its effect on radionuclide solubility," *Chemosphere,* vol. 66, pp. 672-676, 2007.

[3]   K. M. Krupka and R. J. Serne, "Geochemical Factors Affecting the Behaviour of Antimony, Cobalt, Europium, Technetium and Uranium in Vadose sediments.," Pacific Northwest National Laboratory2002.

[4]   J. He, M. Wei, B. Li, Y. Kang, D. G. Evans, and X. Duan, "Preparation of Layered Double Hydroxides," *Structure and Bonding,* vol. 119, pp. 89-119, 2006.

[5]   I. Rousselot, C. Taviot-Guého, F. Leroux, P. Léone, P. Palvadeau, and J.-P. Besse, "Insights on the Structural Chemistry of Hydrocalumite and Hydrotalcite-like Materials: Investigation of the Series $Ca_2M^{3+}(OH)_6Cl \cdot 2H_2O$ (M^{3+}: Al^{3+}, Ga^{3+}, Fe^{3+}, and Sc^{3+}) by X-Ray Powder Diffraction," *Journal of Solid State Chemistry,* vol. 167, pp. 137-144, 2002.

[6]   L. Raki, J. J. Beaudoin, and L. Mitchell, "Layered double hydroxide-like materials: nanocomposites for use in concrete," *Cement and Concrete Research,* vol. 34, pp. 1717-1724, 2004.

[7]   G. Renaudin, J. P. Rapin, B. Humbert, and M. François, "Thermal behaviour of the nitrated AFm phase $Ca_4Al_2(OH)_{12}(NO_3)_2$ - $4H_2O$ and structure determination of the intermediate hydrate $Ca_4Al_2(OH)_{12}(NO_3)_2$ - $2H_2O$," *Cement and Concrete Research,* vol. 30, pp. 307-314, 2000.

[8]   M. J. d. Exter, S. Neumann, and T. Tomasberge, "Immobilization and Behavior of Technetium in a Magnesium Titanate Matrix for Final Disposal," *Materials Research Society Symposium Proceedings,* vol. 932, 2006.

Mater. Res. Soc. Symp. Proc. Vol. 1215 © 2010 Materials Research Society   1215-V11-05

Low Temperature Synthesis of Silicon Carbide Inert Matrix Fuel (IMF)

Chunghao Shih[1], James Tulenko[2] and Ronald Baney[1]
[1]Department of Materials Science and Engineering, University of Florida, Gainesville FL 32611, U.S.A.
[2] Department of Nuclear and Radiological Engineering, University of Florida, Gainesville FL 32611, U.S.A.

ABSTRACT

A process for the synthesis of silicon carbide (SiC) inert matrix fuels at a low temperature (1050 °C) is reported which utilized a liquid polymer precursor. As the polymer content increased, the theoretical density of the pellet at first increased and then reached a plateau. From the onset of the plateau, the packing of the one micron SiC particles in the green body was determined to be 64-68% at 600 MPa pressing pressure. As expected, mixing coarse and fine SiC particles gave a higher pellet density. The maximum density achieved was 80% of the theoretical density. Mercury porosimetry showed that the largest pore size was around 10% of the largest particle sizes present in the green body. SEM images showed that ceria, which was selected as a surrogate for PuO_2 in the present study, was well distributed.

INTRODUCTION

More than 1400 metric tones of plutonium have been created around the world during the past 50 years [1]. In addition to Pu, minor actinides such as Np, Am and Cm with very long half-lives are generated in reactors. A strategy for reducing the amount of both Pu and the minor actinides is to transmute the actinides in a reactor. Mixed oxide fuel (MOX) is the current fuel for plutonium disposition. MOX is a mixture of UO_2 and PuO_2. It contains ^{238}U which is a fertile material that absorbs neutrons and transmute into fissionable ^{239}Pu. A more efficient way would be to replace the MOX by a neutron transparent matrix which contains less or no fertile materials. Thus, the concept of an inert matrix fuel (IMF) has been proposed.

Inert matrix materials are selected for optimum neutronic performance based on the following criteria: low neutron absorption cross section, high melting point, high thermal conductivity, good irradiation stability, and good chemical stability with cladding material and coolant. The properties of silicon carbide (SiC) make it a very promising candidate as an inert matrix material. The thermal conductivity of SiC is 30 W/m-K at 1000 °C [2], which is about 10 times higher than MOX. SiC also has good irradiation stability and good chemical stability to air and hot water [2]. Good mechanical properties normally associated with SiC monolithic parts are not required in SiC IMF's because only sufficient mechanical properties to assemble the fuel are required.

Pre-ceramic polymer precursors have been used as sintering aids to produce SiC [3]. This process offers a route to synthesize SiC at lower temperatures which is desirable for SiC IMF's to avoid reactions between SiC and PuO_2 [4,5] and to avoid the loss of some minor actinides due to their high vapor pressures at higher temperatures. In this paper, we report on a low temperature (1050 °C) synthesis of SiC IMF's with the use of SiC powder and a SiC polymer precursor. The effects of the polymer content, the cold pressing pressure and the particles size composition on the final pellet densities are reported. The pore size distribution is also investigated.

EXPERIMENTAL DETAILS

Materials

Three types of β-SiC powders were used in this study: 1) a coarse powder with a nominal size of 16.9 microns (from Superior Graphite), 2) submicron size powder with nominal size of 0.6 micron (from Superior Graphite), and 3) SiC powder with nominal size of 1 micron (from Alfa Aesar). Since Pu materials are highly toxic and radioactive, cerium oxide (CeO_2) was used as a surrogate for PuO_2 [6]. CeO_2 (from Alfa Aesar) with nominal size 70-100 nm was used in this study. A liquid SiC polymer precursor, SMP-10, manufacturer by Starfire Systems Inc. was used. SMP-10 is an allyl-hydrido-poly-carbo-silane, which yields a 1:1 Si:C ratio and with H as the only other component.

Pellet synthesis and characterization

A conventional pellet fabrication route was used which includes powder preparation, pressing and sintering. Initially, SiC powder with various composition and ceria were mixed with various amounts of polymer precursor in a mortar and pestle for five minutes. The mixture (slurry) was then put into a 13 mm die and uniaxially pressed at various pressures at room temperature for 10 minutes to form green bodies. The green bodies were then sintered in a tube furnace up to 1050 °C under ultra high purity argon. The sintering temperature profile suggested by Starfire Systems Inc. was followed [7]. The weight of the sintered pellet was determined to an

88

accuracy of 0.1 mg. A caliper was used to determine the bulk volume. The difference in the pellet volume determined using a caliper and the conventional Archimedes's method was trivial.

The effects of the polymer precursor content, the cold pressing pressure and the mixing of different SiC particle sizes on the final pellet density after sintering were investigated. The pore size distributions were investigated using a mercury porosimeter (Quantachrome Autoscan 60). Scanning electron microscopy (SEM) was used to observe the cross section of the pellets.

RESULTS AND DISCUSSION

Effect of polymer content

One micron SiC powder was used to study the effect of the polymer content and the pressing pressure on the final density of the pellets. The percent theoretical densities (%TD) after sintering as a function of the polymer weight content in the green body and the cold pressing pressure are shown in Figure 1. Data scattering was negligible. Moreover, the density reached a plateau at approximately 72% TD. When the polymer content was increased, the polymer filled the pores and the excess polymer was squeezed out. This is shown on Figure 1 as dash lines.

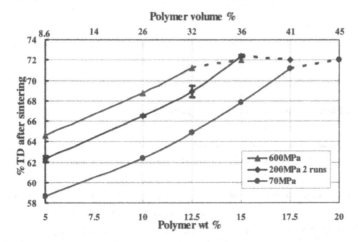

Figure 1. %TD as a function of polymer weight % in the green body and cold pressing pressure

However, during the sintering, the polymer lost 35% of its original weight, and the density changed from 1.0 to 2.4 g/cc [7], which means 73% of the volume originally occupied by polymer precursor became pores. From theoretical calculations, a green body with 15 weight%

polymer precursor and without any pores would sinter into a 73% TD pellet. This fits the experimental data extremely well. From the onset of the green body saturation with the polymer, the packing of the SiC one micron particles in green body was determined to be 64-68% at 600MPa.

Effect of bimodal mixtures

SiC powders with 16.9 micron (coarse) and 0.6 micron (fine) nominal particle sizes were chosen to study the effect of mixing particle sizes variations on the pellet density. The polymer precursor was fixed at 10% by weight and the CeO_2 was fixed at 5% by weight. Various SiC particle compositions were tested with two cold pressing pressures and the results are shown in Figure 2.

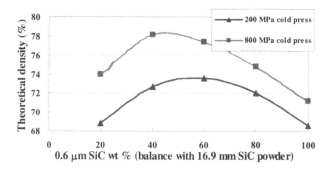

Figure 2. %TD as a function of particle size ratios and pressing pressure.

Mixing of the coarse and the fine SiC particles strongly affects the particle packing. The cold pressing pressure also played an important role. The maximum density achieved was 80% TD which occurred at 60% coarse powder and 40% fine powder. Figure 3 shows the pore size distributions and the cumulative pore volume of six pellets with different SiC particle size compositions and two cold pressing pressures. The pore size distribution was strongly influenced by the composition of particle size. The largest pore size was around 10% of the largest particle size present in the green body. Increasing the cold pressing pressure eliminated some large pores but had little effect on the small pores.

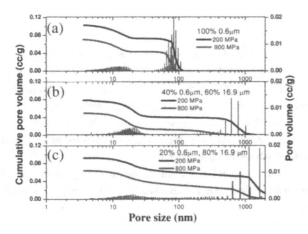

Figure 3. Pore size distribution and cumulative pore volume of pellets with different compositions and pressing pressures.

Figure 4 shows the cross section SEM images (back scattered) of a pellet with 10 weight % polymer precursor and 5 weight % ceria. The SiC pellet was composed of 40% fine particles and 60% coarse particles. It was pressed at 200 MPa. Figure 4 (a) shows that the ceria (white spots) was well distributed. Some agglomerates of 0.6 μm SiC particles with size in the 100 μm range can also be seen (black arrow). It was difficult to break all the agglomerates by a mortar and pestle. A ball mill mixing method is now under investigation. Figure 4 (b) shows a more detailed view of the same pellet. Agglomerates of the small particles can be seen on the right side. Some 16.9 μm SiC particles are shown on the left side with ceria particles packed between SiC grains. The agglomerates size of the ceria is about 1-3 μm in diameter.

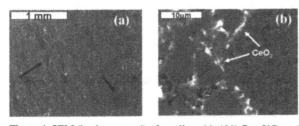

Figure 4. SEM (back scattered) of a pellet with 40% fine SiC particles and 60% coarse SiC particles

CONCLUSIONS

SiC IMFs can be synthesized with the polymer precursor route at a low temperature (1050 °C) which would solve the problem of reactions between SiC and PuO_2 at the higher temperatures required for the conventional SiC sintering. The density and pore size distributions can also be tailored by varying the SiC particle size ratios, polymer content and cold pressing pressure. Furthermore, the ceria (PuO_2 surrogate) was shown to be well distributed. Agglomerates from the 0.6 micron SiC powder were observed in the synthesized pellets. Better mixing methods should be utilized to break the agglomerates.

For IMF, the minimum density required would only be determined by the mechanical strength requirements for handling and operating. The achieved 80% theoretical density should be adequate for IMF applications.

ACKNOWLEDGMENTS

This work is supported by the US Department of Energy under contract number DE-FG07-07ID14775

REFERENCES

[1]   R. C. Ewing, *Earth and Planetary Science Letters,* vol. 229, pp. 165-181, Jan 2005.

[2]   R. A. Verrall, M. D. Vlajic, and V. D. Krstic, *Journal of Nuclear Materials,* vol. 274, pp. 54-60, 1999.

[3]   E. Bouillon, F. Langlais, R. Pailler, R. Naslain, F. Cruege, P. V. Huong, J. C. Sarthou, A. Delpuech, C. Laffon, P. Lagarde, M. Monthioux, and A. Oberlin, *Journal of Materials Science,* vol. 26, pp. 1333-1345, Mar 1991.

[4]   Z. Pan, H. J. Seifert, and O. Fabrichnaya, in *214th ECS Meeting,* Honolulu, HI, USA, 2008, pp. 65-80.

[5]   K. H. Sarma, J. Fourcade, S. G. Lee, and A. A. Solomon, *Journal of Nuclear Materials,* vol. 352, pp. 324-333, 2006.

[6]   H. S. Kim, C. Y. Joung, B. H. Lee, J. Y. Oh, Y. H. Koo, and P. Heimgartner, *Journal of Nuclear Materials,* vol. 378, pp. 98-104, Aug 2008.

[7]   Ed Bongio, Starfire Systems Inc. (private communication)

Structural Complexity in Advanced Nuclear Fuels

Mater. Res. Soc. Symp. Proc. Vol. 1215 © 2010 Materials Research Society   1215-V12-04

Effect of americium and simulated fission products addition on oxygen potential of uranium-plutonium mixed oxide fuels

Kosuke Tanaka[1], Masahiko Osaka[1], Ken Kurosaki[2], Hiroaki Muta[2], Masayoshi Uno[3] and Shinsuke Yamanaka[2, 3]
[1] Japan Atomic Energy Agency, 4002 Narita-cho, Oarai-machi, Higashiibaraki-gun, Ibaraki, 311-1393, Japan
[2] Division of Sustainable Energy and Environmental Engineering, Graduate School of Engineering, Osaka University, 2-1 Yamadaoka, Suita, Osaka, 565-0871, Japan
[3] Research Institute of Nuclear Engineering, Fukui University, 3-9-1 Bunkyo, Fukui, 910-8507, Japan.

ABSTRACT

The oxygen potentials at 1273 K of mixed oxide (MOX) fuels with Am and 26 kinds of fission product elements (FPs), simulating low-decontaminated MOX fuel and high burn-up of up to 250 GWd/t, have been measured by using thermogravimetric analysis (TGA). The oxygen potentials for simulated low-decontaminated MOX fuels were higher than the fuels without FPs and increased with increasing simulated burn-up.

INTRODUCTION

Low-decontaminated mixed oxide (MOX) fuel, which contains several percent of minor actinides (MAs) and fission products (FPs), is a promising candidate for a closed nuclear fuel cycle system based on a fast reactor [1]. Extending burn-up of the fuel has been identified as a practical means of improving the economics of the system. In high burn-up oxide fuels, some FPs dissolve in the fuel matrix and others form oxide or metallic precipitates, which will affect the physical and chemical properties of the fuels [2]. It is, therefore, crucial to understand the effect of FPs accumulation on the fuel performance.

Among the thermodynamic properties, the oxygen potential is one of the concerns for oxide fuels [3]. The oxygen potential greatly affects thermochemical and thermophysical properties of non-stoichiometric nuclear fuel via variation of the oxygen-to-metal (O/M) ratio. This means that effects of the oxygen potential should be considered with great care for the thermal design of the fuel. The quality of fabricated fuel pellets is also influenced by the oxygen potential in their sintering atmosphere [4, 5]. Besides, it is an indispensable property for interpreting migration behavior of the fuel constituent elements under a temperature gradient, especially in the initial stage of the irradiation. During the irradiation, effects of the oxygen potential become more critical as it has direct influences on the oxidation of the cladding material, which may eventually lead to cladding failure due to the reduction of the effective load-bearing thickness. Oxygen potential is also closely related to various aspects of the fuel during the long term irradiation.

There are some experimental data for oxygen potentials of the oxides containing americium (Am) which is representative of MAs. Chikalla and Eyring [6] measured the oxygen potentials of AmO_{2-x} as a function of the O/M ratio at several temperatures. Their study showed that AmO_{2-x} has very high oxygen potentials, compared with other related actinide oxides such as UO_{2+x}, PuO_{2-x} and $(U,Pu)O_{2\pm x}$, where x represents the deviation from stoichiometry, O/M = 2. Regarding the oxygen potentials of the oxides including Am, Bartscher and Sari [7] investigated the dependences of the oxygen potentials of $(U_{0.5}Am_{0.5})O_{2\pm x}$ on O/M ratios and temperatures, and found that the oxygen potentials were much lower in $(U_{0.5}Am_{0.5})O_{2-x}$ than in AmO_{2-x}, but were still much higher than in $(U,Pu)O_{2-x}$. Osaka et al. [8, 9] measured the oxygen potentials of non-stoichiometric solid solution by thermogravimetric analysis as a function of oxygen-to-metal ratio at 1123, 1273 and 1423 K. They found that the addition of AmO_2, even in an amount as low as 4.5%, to $(U,Pu)O_{2-x}$ significantly increased the oxygen potentials. Recently Kato et al. [10] reported the oxygen potentials of MOX fuel with 2% Am. Regarding the effect of FPs accumulation on the variation of oxygen potentials, the experimental data are restricted to one study by Une and Oguma [11] on LWR fuel and one by Woodley [12] on MOX fuel, both up to 10 at.% burn-up. To the best of the authors' knowledge, no research has as yet been carried out to investigate the effect of Am and FP addition on the oxygen potentials of MOX fuel. The objective of this study was to contribute to the development of low-decontaminated MOX fuel by determining oxygen potential-related behavior of the MOX fuels including Am and FPs.

EXPERIMENTAL

Sample preparation and characterization

MOX fuel containing 3%Am and 1.5 % FPs was defined as a low-decontaminated MOX fuel in this study. Based on the composition of the fuel, three simulated compositions were made, representing the burn-ups of 0, 150 and 250 GWd/t. The ORIGEN2 code was used to determine the fuel compositions.

Table1. Composition (as metal atomic fraction) of mixed powders

Metal	Burn-up (GWd/t) 0	150	250		0	150	250		0	150	250
U	66.500	49.324	40.206	Rh	0.005	0.844	1.205	Eu	0.106	0.193	0.212
Pu	29.000	21.133	16.843	Pd	0.013	2.287	3.894	Gd	0.175	0.273	0.341
Am	3.000	1.768	1.235	Te	0.003	0.523	0.798	Tb	0.010	0.013	0.015
Rb	0.001	0.256	0.384	I	0.001	0.295	0.416	Dy	0.010	0.020	0.026
Sr	0.003	0.605	0.894	Cs	0.015	3.165	4.775	Ho	2.3E-04	2.5E-04	3.3E-04
Y	0.594	0.938	1.033	Ba	0.006	1.098	1.761	Er	3.2E-04	6.2E-04	7.9E-04
Zr	0.015	3.026	4.567	La	0.005	0.921	1.398	Tm	1.0E-06	1.0E-06	1.0E-06
Mo	0.017	3.334	5.129	Ce	0.008	1.969	2.852	Yb	1.0E-06	1.0E-06	1.0E-06
Ru	0.015	3.603	5.342	Pr	0.004	0.803	1.213	Lu	1.0E-06	1.0E-06	1.0E-06
				Nd	0.024	2.422	3.841	SUM	100	100	100
				Sm	0.468	1.185	1.617				

The samples were prepared by a conventional powder metallurgical route in a glove box. Powders of UO_2, PuO_2 and Am-containing PuO_2 powders were used as raw materials. Detailed

96

characterization of the raw powders has been reported in the literature [13]. These powders, together with 26 kinds of simulated FP elements were weighed by using an electronic balance in order to adjust the amount to the predetermined weight ratio. Table 1 gives the composition of the mixed powders. The weighed powders with a small amount of polyvinyl alcohol (PVA) binder were thoroughly mixed in an agate mortar with a pestle in acetone medium. The mixed powder was then compacted into a columnar pellet by a uni-axial pressing unit and sintered in a reducing atmosphere. Eventually all the samples were heat treated at around -250 kJ/mol in the reducing atmosphere. There were neither visible cracks nor any failures in any sintered and heat treated body. 3%Am-MOX fuel sample as a reference was also prepared by the same fabrication procedure.

The sintered and heat treated pellet samples were characterized by X-ray diffraction (XRD) analysis using a Rigaku model RAD-C system, by ceramographic observations using an optical microscope (TELATOM, Reichert) and by electron probe micro analysis (EPMA) using CAMECA model SX-100R.

Thermogravimetric analysis

After crushing the pellets, about 150 mg of each specimen was subjected to TGA. The TGA was carried out at 1273 K, using a Rigaku model TG-8120 connected with a gas supply system. Each specimen was loaded into an alumina capsule and placed in the TGA apparatus. For reference, an alpha-alumina sample was loaded into another alumina capsule and also placed in the apparatus. Oxygen partial pressure was adjusted at a constant temperature. The adjustment was done by gas equilibria according to target oxygen partial pressures by changing the ratio of H_2O/H_2. The inlet and outlet oxygen partial pressures were measured with stabilized zirconia oxygen sensors. Microgram order weight changes in the specimen were continuously monitored while changing the oxygen partial pressure at the pre-determined temperature step by step. The oxygen potentials were calculated by using Eq. (1):

$$\Delta G_{O_2} = RT \ln p_{O_2} \tag{1}$$

where ΔG_{O_2} is the oxygen potential, R is the gas constant, T is the temperature and p_{O_2} is the oxygen partial pressure, which is derived from the ratio of oxygen partial pressure to its standard state, 0.101 MPa. The O/M ratios at various oxygen partial pressures and temperatures were calculated from the weight changes relative to the stoichiometry, O/M = 2. Estimated errors of the oxygen potential and O/M ratio were ±10 kJ/mol and ±0.002, respectively.

RESULTS AND DISCUSSION

Characterization of the samples

The ceramography and EPMA results revealed that the U, Pu and Am together with the rare earth and other oxides were dissolved in the matrix uniformly and small noble metallic and gray phase oxide precipitates were distributed throughout the pellets. The size and the number of these precipitations increase with increasing the simulated burn-ups. X-ray diffraction patterns with Cu-K$_\alpha$ radiation of the samples, however, showed single phase fluorite structures. No sign of any other phase formation was found by XRD analysis. A width of a peak was almost same as that of (U, Pu)O$_2$ fuel and the peak was split into K$_{\alpha1}$ and K$_{\alpha2}$. This indicated that the matrix phases of these samples were almost completely homogeneous solid solution. The lattice parameters decreased with increasing FPs content, the same as reported in the literature [14-16]. The lattice parameters which were calculated by ion radii of the metal and oxygen together with the ionic fraction of cation were in good agreement with measured ones. Therefore, the O/M ratio of the samples after the final heat treatment was nearly equal to 2.00. It was assumed that the equilibrium at the oxygen potential around -250kJ/mol, which was the final heat treatment condition, corresponded to O/M ratio of 2.00 in this study.

Oxygen potentials

The oxygen potential is a function of temperature and O/M ratio of the oxide. It is, therefore, important to prepare the oxygen potential data as a function of them. Fig 1 shows the oxygen potentials obtained by the present measurements at 1273 K for 3%Am-MOX and a sample which corresponded to 0 GWd/t of the simulated low-decontaminated MOX fuel. ΔG_{O_2} increased with increasing O/M ratio. Furthermore, a particularly steep increase of ΔG_{O_2} appeared around the stoichiometry, O/M = 2.00, according to the definition made in the previous section. As shown in this figure, ΔG_{O_2} of the sample which corresponded to 0 GWd/t was slightly higher than 3%Am-MOX. This implied that, even for a small FP addition of 1.5%, careful consideration would be required for evaluating the impact on the thermal performance of low-decontaminated MOX fuels.

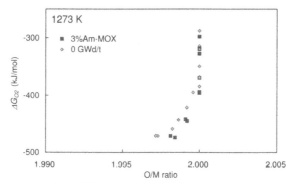

Figure 1. Oxygen potentials at 1273 K for 3%Am-MOX and a sample which corresponded to 0 GWd/t of the simulated low-decontaminated MOX fuel. ΔG_{O_2} of the sample which corresponded to 0 GWd/t was slightly higher than 3%Am-MOX.

Figure. 2 shows ΔG_{O_2} of samples which corresponded to 0, 150 and 250 GWd/t of low-decontaminated MOX fuel as a function of O/M ratio at 1273 K. The effect of the simulated fission products on the ΔG_{O_2} value was clearly seen. This tendency was the same as that seen in the data measured by Une and Oguma [11] on simulated LWR fuel, and by Woodley [12] on simulated MOX fuel with up to 10 at % burn-up.

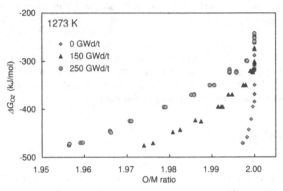

Figure 2. ΔG_{O_2} of samples which corresponded to 0, 150 and 250 GWd/t of low-decontaminated MOX fuel as a function of O/M ratio at 1273 K. The effect of the simulated fission products on the ΔG_{O_2} value was clearly seen.

In Figure. 3, the ΔG_{O_2} values are plotted as a function of simulated burn-up at O/M ratio of 1.998. The ΔG_{O_2} increased positively with simulated burn-up up to 250 GWd/t. The increase rate of ΔG_{O_2} up to 150 GWd/t was almost the same as Woodley obtained [12]. On going from the 150 GWd/t samples to 250 GWd/t samples, this slope decreased.

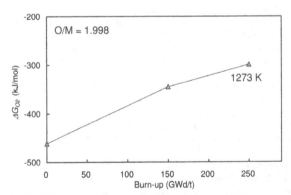

Figure 3. Variation in ΔG_{O_2} of samples of low-decontaminated MOX fuel with simulated burn-up at O/M ratio of 1.998. The ΔG_{O_2} increased with simulated burn-up up to 250 GWd/t.

CONCLUSIONS

In order to investigate the effect of Mas and FP addition on the oxygen potential of MOX fuels, thermogravimetric analysis (TGA) were carried out. MOX fuels with Am and 26 kinds of fission product elements (FPs), simulating low-decontaminated MOX fuel and high burn-up to 250 GWd/t, were prepared by a conventional powder metallurgical route in a glove box. The oxygen potentials for simulating low-decontaminated MOX fuels were higher than the fuels without FPs and increased with increasing simulated burn-up. These results would be useful for evaluating the performance of the low-decontaminated MOX fuels from the initial stage of irradiation to the targeted high burn-up in fast reactors.

ACKNOWLEDGMENTS

The authors would like to express their appreciation to Messrs. Hirosawa and Miwa of Alpha-Gamma Section (AGS) of Oarai Research and Development Center of JAEA for their technical support. The authors also greatly appreciate the cooperation of Messrs. Sekine (Nuclear Technology and Engineering Corporation) and Seki (Inspection Development Company). The present study includes the results of "Study on the physical properties of nuclear fuels with multi phase system: Toward establishment of the closed cycle system with low-decontaminated oxide fuel" entrusted to Osaka University by the Ministry of Education, Culture, Sports, Science and Technology of Japan (MEXT).

REFERENCES

1. T. Namekawa et al., Proc. Int. Conf. GLOBAL 2005, Tsukuba, Japan, Oct. 9-13, 2005, paper No. 424.
2. H. Kleykamp, J. Nucl. Mater., **131**, 229 (1985) .
3. D. R. Olander, Fundamental Aspects of Nuclear Reactor Fuel Elements, TID-26711-P1, 1970.
4. R. Yuda, K and Une, J. Nucl. Mater. **178**, 195(1991).
5. S. Miwa et al., Recent Advances in Actinide Science, Manchester, July 4-8, 2005, eds. I. May, N. D. Bryan and R. Alvarez, RSC Publishing: pp. 400-402, 2006.
6. T. D. Chikalla and L. Eyring, J. Inorg. Nucl. Chem. **29**, 2281 (1967).
7. W. Bartscher and C. Sari, J. Nucl. Mater. **118**, 220 (1983).
8. M. Osaka et al., J. Alloys Com., **397**, 110 (2005).
9. M. Osaka et al., J. Alloys Com., **428**, 355 (2007).
10. M. Kato et al., J. Nucl. Mater., **385**, 419 (2009).
11. K. Une and M. Oguma, J. Nucl. Sci. Technol., **20**, 844 (1983)
12. R.E. Woodley, J. Nucl. Mater., **74**, 290 (1978)
13. H. Yoshimochi et al., J. Nucl. Sci. Technol. **41**, 850 (2004).
14. P.G. Lucta et al., J. Nucl. Mater., **188**, 198 (1992).
15. K. Une and M. Oguma, J. Nucl. Mater., **131**, 88 (1985).
16. S. Ishimoto et al., J. Nucl. Sci. Technol., **31**, 796 (1994).

Modeling Complex Materials II

Mater. Res. Soc. Symp. Proc. Vol. 1215 © 2010 Materials Research Society   1215-V13-03

How to Simulate the Microstructure Induced by a Nuclear Reactor with an Ion Beam Facility: DART

Laurence Luneville[1], David Simeone[2], Gianguido Baldinozzi [3], Dominique Gosset[2], Yves Serruys[4]

[1]CEA/DEN/DANS/DM2S/SERMA/LLPR, Matériaux Fonctionnels pour l'Energie, Equipe Mixte CEA-CNRS-ECP, CE Saclay, Gif sur Yvette, 91191, France
[2]CEA/DEN/DANS/DMN/SRMA/LA2M, Matériaux Fonctionnels pour l'Energie, Equipe Mixte CEA-CNRS-ECP, CE Saclay, Gif sur Yvette, 91191, France
[3]CNRS, Matériaux Fonctionnels pour l'Energie, Equipe Mixte CEA-CNRS-ECP, SPMS, ECP, Chatenay-Malabry, 92292, France
[4]CEA/DEN/DANS/DMN/SRMP, CE Saclay, Gif sur Yvette, 91191, France

ABSTRACT

Even if the Binary Collision Approximation (BCA) does not take into account relaxation processes at the end of the displacement cascade, the amount of displaced atoms calculated within this framework can be used to compare damages induced by irradiation at different facilities like pressurized water reactors (PWR), fast breeder reactors (FBR), high temperature reactors (HTR) and ion beam facilities on a defined material. In this paper, a formalism is presented to evaluate the displacement cross-sections pointing out the effect of the anisotropy of nuclear reactions. From this formalism, the impact of fast neutrons (with a kinetic energy E_n larger than 1 MeV) is accurately described. This point allows calculating accurately the displacement per atom rates as well as primary and weighted recoil spectra. Such spectra provide useful information to select masses and energies of ions to perform realistic experiments at ion beam facilities.

INTRODUCTION

During more than 50 years, many experimental works were performed to study radiation damage in materials. Submitted to important neutron fluxes, materials in nuclear plants exhibit unusual microstructures and subsequent properties [1]. Many works in 1980s have clearly shown a large shift of the brittle-ductile transition temperature in steel under irradiation [2]. Looking back into history, we find prominent names like Bohr, Fermi and Bethe associated with the early development of this field of research. Many works were devoted to capture key parameters responsible for the unexpected microstructures of solids observed under irradiation. Even if primary damage due to neutron - atoms interactions are produced at the atomic scale, the spatially inhomogeneous distribution of these primary defects, their diffusion and their clustering encompass different length scales giving rise to complex microstructures. Many works were devoted to study the slowing down of ions in matter [3-7]. However, comparatively few works in the material field were performed to quantify the primary damages induced by neutrons in solids [8-11]. This point explains why ion beam experiments remain the most efficient tool to study the structural stability of solids under irradiation [3-7]. From this short introduction, a question naturally arises: can we select peculiar ions to simulate the microstructure induced by neutron irradiations occurring in nuclear plants?

The answer to this question is quite complex. The neutron-atom cross sections are 8 orders of magnitude smaller than the ion-atom cross sections for iron irradiated in a core of a Pressurized Water Reactor at the middle life and irradiated by 2 MeV Au^+ ions at room temperature. This point implies that:

- The displacement per atom rate due to typical neutron fluxes (core of a Pressurized Water Reactor at the middle life) is thus about 3 orders of magnitude smaller than the displacement per atom rate induced by ion beams (2 MeV Au^+ ions and a flux of $10^{12}cm^{-2}s^{-1}$ irradiated at room temperature). Neglecting the diffusion of point defects at low temperature, a correct scaling of the irradiation time in particle accelerators can easily account for this effect in a first approximation. In practice, experimentalists increase the temperature during an ion irradiation to avoid possible overlapping between collisions cascades.

- The neutron mean free path between two collisions in the target (a few centimeters) is larger than the ion mean free path (a few nanometers). The localization of primary defects in the medium should then be drastically different. This remark can be also easily accounted for. As the neutron ion mean free path is large, the characteristic length between two cascades in neutron irradiations is so large that displacement cascades (symbolized by large grey areas on figure 1) can be considered as independent events. Indeed, they never overlap. In ion beam experiments, the penetration depth of incident particles with kinetic energy below a few MeV is of the order of a few nanometers. Then, this analysis points out the fact that both irradiations generate the same localization of primary defects over a few nanometers. .

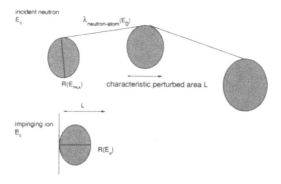

Figure 1. This graph sketches the different length scales associated with the penetration of particles in a medium (top: penetration depth of a neutron in the medium, bottom: penetration depth of an ion in the medium).

Figure 1 summarizes the characteristic lengths associated with the microstructural evolution of solids under ion and neutron irradiations.

As the ion-atom interatomic cross-sections are much more larger than the neutron atoms ones, penetration depths $R(E)$ of ions is of the same order of magnitude as the displacement cascade L produced by neutrons [12]. This point ensures that ion beams can efficiency simulate the evolution of materials irradiated by nuclear plants.

As pointed out on Figure 1, all characteristic lengths depend only on the energy of incidents projectiles. It appears that criteria based only on the energy of recoil atoms would thus give some clues to select the energies and masses of ions to simulate radiation damage occurring in materials in nuclear plants. In this work, we describe in detail a new formalism to take into account different peculiar inelastic neutron-atoms cross-sections in the calculation of the displacement per atom rate and the primary and recoil spectra associated with a defined neutron irradiation. Thanks to this formalism, we establish some criteria to select the mass and the energy of incident ions able to reproduce in metals and ceramics the same microstructures as those created in nuclear facilities.

THEORY

When a material is impacted by a neutron, atoms are displaced from their lattice sites with a large amount of energy [13,14] leading to the appearance of displacement cascades and then to point defects and extended defects at well localized areas. So, a precise description of neutron-atoms inelastic interactions is essential to calculate accurately the amount of energy transferred to primary knocked-on atoms (PKA). The Isotropic Emission Compound Nucleus model (IECN) [15,9], was usually used to perform these calculations. Although the complete angular distribution is used for elastic scattering, for inelastic processes, such as inelastic diffusion or emission of particles, the angular distributions of recoils are assumed isotropic in the center of mass system within this framework. However, recent nuclear evaluations (ENDFB6, JEFF3) [16,17] contain accurate angular distributions of recoils for all neutron-atom interactions. We present in this paper a new formalism to take into account this information. This formalism highlights the impact of this anisotropy on the recoil energy distribution [18,19].

An incident neutron of energy E generates a recoil of energy T for a specific reaction (as for example inelastic scattering). Defining θ as the angle between the incident and the recoil particle, the corresponding differential cross-section for this nuclear reaction is directly obtained from nuclear evaluations in the following way:

$$\sigma(\mu, E, E') = \sigma(E) f(\mu, E, E') / 2\pi \qquad (1)$$

where $\mu = cos\theta$ is the cosine of the angular deflection in the center of mass frame, $\sigma(E)$ is the neutron reaction cross section, E' the energy of the emitted particles and $f(\mu, E, E')$ is the angular emission probability density function. The differential PKA cross-section, $\chi(E,T)$, is then expressed as :

$$\chi(E,T) = \sigma(E) \int_0^{+\infty} dE' \frac{f(\mu, E, E')}{2\pi} \frac{1}{\left|\frac{\partial T}{\partial \mu}\right|}. \qquad (2)$$

105

In the IECN model, $f(\mu,E,E')$ is independent of μ. In many inelastic nuclear reactions, this function is peaked and sharp. From Eq.2, it appears clearly that the shape of $f(\mu,E,E')$ modifies drastically $\chi(E,T)$. Figure 2 displays the evolution of $\chi(E,T)$ as a function of the recoil energy T for a given neutron energy E. The function $\chi(E,T)$ calculated within our formalism exhibits large discrepancies in $\chi(E,T)$ derived from the IECN model. This point highlights the effects on the

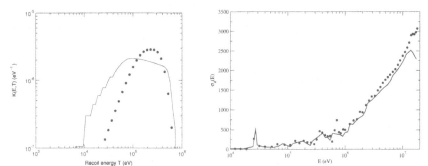

Figure 2. Comparison between the PKA cross-section and the displacement cross-section as a function of the energy of recoils (left) and the energy of incident neutron (right) calculated for a Fe target irradiated under a 14 MeV neutron flux. The large discrepancy between $\chi(E,T)$ derived from our formalism (dots) and calculated within the IECN approximation (full line) appears clearly on figure 2 (left). This discrepancy of $\chi(E,T)$ leads to the large variation of the displacement cross-sections calculated according to our formalism (dots) and within the IECN approximation (full line). This discrepancy is equal to 20% in the high energy domain (above $E_n>1$ MeV) where the neutron spectra exhibit large values in nuclear facilities.

From the calculation of $\chi(E,T)$, it is possible to derive the displacement cross-section $\sigma_d(E)$ according to :

$$\sigma_d(E) = \int_0^{T_{max}} \chi(E,T)\nu(T)dT \qquad (3)$$

where T_{max} is the maximum energy transferred to the recoil atom and $\nu(T)$ is the mean number of displaced atoms calculated within the BCA formalism [10,21,22]. From this equation, it appears clearly that our formalism describes accurately the high energy part of $\sigma_d(E)$ associated with inelastic neutron-atom collisions. The displacement per atom rate, P, is obtained summing the displacement cross-section weighted by the neutron spectrum over the neutron energy:

$$P = \int \sigma_d(E)\phi(E)dE \qquad (4)$$

where $\sigma_d(E)$ is the displacement cross-section and $\phi(E)$ is the neutron spectrum (n cm^{-2}s^{-1} eV^{-1}). As $\sigma_d(E)$ is an increasing function and the neutron spectrum exhibits a sharp shape aroud a few MeV, $\sigma_d(E)$ for $E_n>1$ MeV gives the main contribution to the displacement per atom rate[23]. This point highlights the need for an accurate description of inelastic neutron collisions taking

106

into account the anisotropy of these reactions. From Eq.4, it appears clearly that the dpa rate results from the cumulative effect due to the flux of incident particles and the displacement induced in the solids by each impinging particle. In the sum, the nature of the incident particle is averaged by its cross section (the term $\chi(E,T)$ in Eq.3). From this analysis, it is clear that the dpa rate does not only take into account the nature of the facility, i.e. nature of incident particles and flux, but also the complexity of the irradiated solid, i.e. number of species forming the solid and displacements induced by each PKA in the solid). Despite, the dpa rate is useful to compare different irradiations performed with different facilities. The integration along the energy spectrum of impinging particles erases all information on the energy distribution of PKA and recoils. It is now clearly understood that the variation of the microstructure in irradiated solids depends on this distribution [1-3].

To compare different irradiations performed in ion beam facilities to irradiation in nuclear plants, the primary and recoil spectra need to be calculated. A program called DART has been achieved to provide accurate displacement cross-sections, displacement per atom rates as well as different spectra for a polyatomic solid irradiated by neutrons, ions or electrons. In this program, the number of displaced atoms is computed according to the Lindhard formalism within the BCA. On the other hand, neutron-atoms cross sections were derived from specific library (such as ENDFB6) instead of the classical IECN model. A clear description of this program as well as different comparisons between this program and other codes like SPECOMP or SRIM can be found in some articles [10,18,19].

DISCUSSION

The renewed interest in nuclear energy production is bringing about a renaissance in materials sciences. In order to test new concepts, the structural stability of SiC and Oxides Dispersed Strengthened (ODS) steels need to be studied in great detail. From the knowledge of neutron spectra of generation IV plants, displacement cross-sections and different spectra can be calculated using DART. To simulate the structural stability of ODS under irradiation, the masses and energy of particles (ions or electrons) must be chosen to give similar spectra. Figure 3 displays the evolution of primary and recoil spectra induced by a neutron flux extracted from a FBR reactor, an ion beam facility and an electron flux. Whereas only an irradiation performed with 1.2 MeV Kr ions seems to give the same primary spectrum as a High Temperature Reactor, only an irradiation performed with 600 keV Ar ions gives the same recoil spectrum as an

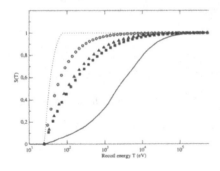
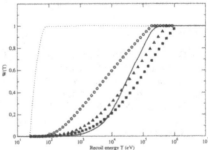

Figure 3. Evolution of the primary (left) and recoil spectra (right) in ODS steel as a function of the recoil energy for different projectiles: 1MeV electrons (dots), 1.2 MeV Krypton ions (squares), 1MeV Helium ions (circles), 600keV Argon ions (triangles) and FBR neutron flux (solid line).

It is important to note that the recoil energy spectrum must be calculated to compare ion and neutron irradiations as the main part of displacements are due to recoils. As pointed out by Eq.5, the number of displaced atoms appears clearly in the calculation of $W(T)$.

$$W(T) = \frac{\int_0^{E_n} dE \int_0^T \chi(E,S)\nu(S)\phi(E)dS}{\int_0^{E_n} dE \int_0^{T_{max}} \chi(E,S)\nu(S)\phi(E)dS} \tag{5}$$

Transmission Electron Microscopy observations performed on ODS steels irradiated at 30 dpa under a FBR neutron flux, and 1.2 MeV Kr and 600 KeV Ar irradiations clearly assess that only similar recoil spectra give the same microstructure in this kind of material [20,24,25].

CONCLUSION

Many works were devoted to study the slowing down of ions in matter to simulate radiation damage produced in nuclear plants. Few works described in detail the energy transfer from neutrons to atoms during neutron-atom collisions. The IECN model is usually used to quantify the impact of neutron-atom collisions. However, the anisotropy of inelastic neutron atom collisions, preponderant in a nuclear plant, is not accurately described in this model. In this paper, we present a formalism to overcome this point. We developped a program (DART) to calculate the displacement per atom rate as well as primary and recoil spectra induced by high energy neutrons in a polyatomic target. This paper highlights the fact that this anysotropy is responsible for a large increase of the displacement cross-section and a drastic modification of primary and recoil spectra. The calculation of accurate primary and recoil spectra allows to select the mass and the energy of ions able to simulate radiation damage induced by neutrons in an ODS steel. The analysis of these spectra reveals that 600 keV Ar ions are able to produce an evolution of the microstructure in this complex material, similar to neutron irradiations in FBR. TEM observations of the irradiated microstructures assess this point.

This work points out that tools are now available to select ions produced in ion beam facilities to simulate radiation damage in solids occurring in nuclear plants. Taking into account the peculiar feature of neutron atom inelastic collisions, the comparison of recoil spectra gives clues to choose the mass and the energy of impinging ions in a realistic way.

REFERENCES

1. G. Martin, P. Bellon, *Solid State Phys.* **50**, 189 (1997).
2. E. Balanzat, S. Bouffard, *Materials under Irradiation,* Trans Tech. Publication (1993).
3. C. Abromeit, *J. of Nucl. Mat.* **216**, 78 (1994).
4. M. Bleiberg and J. Bennet, Int. Conf. on *Radiation Effects In Breeder Reactor Structural Materials*, ASTM Scottsdale (1977).
5. H. Brager and J. Perrin, Int. Conf. on *Effects of Radiation on Materials*, ASTM Scottsdale (1982).

6. F. Garner, N. Packan and A. Kumar, Int. Conf. on *Radiation Induced Changes in Microstructure*, ASTM Seattle (1986).
7. G. Odette, *J. of Nucl. Mat.* **85-86,** (1979).
8. R. Averback, *J. of Nucl. Mat.* **33**, 108 (1971).
9. G. Odette, D. Doiron, *Nuclear technology* **29**, 346 (1976).
10. D. Simeone, O. Hablot, V. Micalet, P. Bellon, Y. Serruys, *J of Nucl. Mat* **246**, 206 (1998).
11. L. Greenwood, R. Smither, SPECTER: Neutron Damage Calculation For Materials Irradiations, *ANL/FPP/TM*-**187** (1985).
12. D. Simeone, L. Luneville, J. P. Both, *Euro. Phys. Let.* **83**, 56002 (2008).
13. W. Meyerhoff, *Elements of Nuclear Physics*, Dunod (1970).
14. K. Kikuchi, M. Kawai, *Nuclear Matter and Nuclear Reactions*, North Holland (1968).
15. L. Greenwood, *J of Nucl. Mat.* **206**, 25 (1994).
16. V. MacLane, ENDF-102 Data Formats for the Evaluated Nuclear Data file ENDF-6, *Cross Section Evaluation Working Group*, BNL-NCS-44945-02/04-Rev (2001).
17. The JEFF3.0 Nuclear Data Libray, *JEFF Report* **19**. NEA. OCDE (2000).
18. D. Simeone, L. Luneville, C. Jouanne, *J. of Nucl. Mat* **353,** 89 (2006).
19. L. Luneville, D. Simeone, D. Gosset, *Nucl. Instr. Meth. B* **250**, 71 (2006).
20. M. Klimiankou, R. Lindau, A. Möslang, *J. of Crystal Growth* **249,** 381 (2003).
21. A. Albermann, D. Lesueur, *Am. Soc. for Tes. and Mat* **19**, 19103 (1989).
22. J. Lindhard, V. Nielsen, M. Schraff, Kgl. *Dan. Vid. Mat. Fys. Medd* **36**, 1 (1968).
23. R. Averback, T. Diaz de la Rubia, *Solid State Phys.* **50**, 189 (1997).
24. I. Monnet, Ph. D. (2000).
25. C. Chen, J. Sun, Y. Xu, *J. of Nucl. Mat* **283-287**, 1011 (2000).

Poster Session: Advanced Materials for Nuclear Energy

Mater. Res. Soc. Symp. Proc. Vol. 1215 © 2010 Materials Research Society 1215-V16-09

Quantification of Trace and Ultra-trace Elements in Nuclear Grade Manufactured Graphites by Fast-Flow Glow Discharge Mass Spectrometry and by Inductively Coupled Plasma - Mass Spectrometry after Microwave - Induced Combustion Digestions

Xinwei Wang, Gaurav Bhagat, Kevin O'Brien and Karol Putyera
Evans Analytical Group - New York, 6707 Brooklawn Parkway, Syracuse, NY 13211, U.S.A.

ABSTRACT

Fast-Flow Glow Discharge Mass Spectrometry (FF-GDMS) and Inductively Coupled Plasma – Mass Spectrometry after Microwave – Induced Combustion Digestions (MIC-ICP-MS) methods were developed for the determination of mg/kg- and μg/kg-level B, Mg, Al, Ca, Ti, V, Cr, Mn, Fe, Co, Ni, Cu, Zn, Zr, Mo, Sb, W and Pb in nuclear-grade graphite. Consistent results have been achieved in determining trace elements like B, Ti, Cr, Mn, Zr, Sb and Pb by both methods, which vary mostly less than ±30%, and are in line with the manufacturer reference values. On Mg, Al, Fe, Co, Zn, Mo and W, FF-GDMS analyses also show good agreement with the manufacturer's data. Continuing efforts in identifying source of interference, which has limited the MIC-ICP-MS analysis of these elements, is currently underway.

INTRODUCTION

"Nuclear"-grade graphite is the core structural material for the high-temperature gas-cooled (usually helium) reactor systems. Its performance and lifetime not only are closely related to the irradiation environment but also are dramatically affected by the specifics of the particular graphite: manufacturing process, graphitization temperature, composition (amount of coke, filler, etc., depending on where it was mined), and so on[1].

It is well documented that chemical and elemental impurities catalyse variety of reactions in graphite. The nuclear industry has come to realization that specific elemental impurities, even at trace levels (mass fractions on the mg/kg levels) could lead to damage of the structural integrity and thus affecting the long-term performance of graphite-made components[2,3]. For example, trace elements, such as Fe, Ni and Al accelerate the corrosion of graphite; Fe, Al, Cu and Ca promote thermal and radiolytic oxidation at the presence of trace CO_2/CO in helium or in the event of air/water ingression [4]; B, Si and Ti enhance the sublimation of graphite [5]; B, Gd, Sm and Cd capture neutrons [6,7], whereas Li, Si, Al and others form undesirable artificial radioactive isotopes. There are suggested limits of elemental impurities identified in ASTM D7219, which are detrimental for isotropic & near-isotropic nuclear graphite (Table I). These impurities may originate from the precursors (e.g., coke, pitch, synthetic polymers), and/or be introduced either during graphitization processes and/or in the post-graphitization processes.

Great efforts have been made on development of analytical protocols for measuring trace level concentrations in graphite [8], much of which were directed on solid sampling methods, like laser ablation inductively coupled plasma-atomic emission spectroscopy [9], solid sampling electrothermaovaporization atomic absorption and/or atomic emission spectrometry using the boat technique [10], instrumental neutron activation analysis (INAA) [11], glow discharge atomic absorption spectrometry [12] and X-ray fluorescence spectroscopy (XRF). INAA is important for certifying trace analyte content, but is not easily accessible for routine analysis. All other solid sampling methods require strict standardization procedures using certified reference materials. Unfortunately, availability of certified graphite based reference materials is rather limited, and solid forms are virtually non-existent for high purity and/or nuclear grades. Consequently, synthetic laboratory standards or liquid alternative approaches [13] were developed to circumvent this issue. Wet digestion of graphite with acid mixtures has been problematic, regardless if it is done in open or pressurized closed vessels, in microwave ovens, or by combustions. The situation exacerbates regarding nuclear grade graphite, for the flip side of which being able to endure the stringy working conditions also means it is extremely difficult, if not impossible, to break down by common laboratory means. As a result, solution based analyses of graphite samples have been subject to great uncertainties. In general, the industrial and academic communities are still working on preferred test methods, especially for survey analyses of purified graphite grades and specifically manufactured carbons.

Table I: Elemental specification for isotropic & near-isotropic nuclear graphites according to ASTM D7219 [Oakridge National Laboratory]

Impurity Category*	Element	Suggested Limit mg/kg	Remark
OPC	Al	< 10	Suggested based on typical maximum observed values and possible contribution to oxidation
OPC	Ba	< 10	Suggested based on maximum observed values and possible contribution to oxidation
NAI	B	< 1.0	Strong neutron absorber; Difficult to remove from graphite Suggested value is well above maximum observed values for purified grades
OPC	Ca	< 10	Suggested based on maximum observed values and contribution to catalytic oxidation
ARI/NAI	Cd	< 0.05	Suggested based on maximum observed values and contribution to neutron absorption & activity
ARI	Ce	< 0.05	Suggested; Not routinely analyzed
ARI /MCRI	Cl	< 5	Suggested based on maximum observed value for electrographite in fuel matrix graphite Not routinely analyzed - analysis may be problematic and prone to scatter
OPC	Cu	< 10	Based on maximum observed values

* ARI – activation relevant impurities; FFE – fissile/fissionable element; MCRI – Metallic corrosion relevant impurities; NAI – neutron absorbing impurities; OPC – oxidation promoting catalysts.

Glow discharge mass spectrometry is among the most sensitive direct sampling analytical methods today for full survey analysis of solids.[14] The direct sampling is accomplished by atomization / plasma sputtering of the sample in the glow discharge ion sources. The released atoms or atom clusters are subsequently ionized in the discharge, extracted and accelerated into the mass analyzer. Since atomization

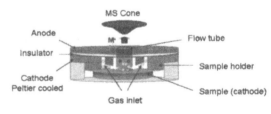

Figure 1: Schematic FF-GD ion source.

and ionization are separated in space and time, the mass spectrometer is sampling from a normalizing gas matrix in which the analytes lost their chemical memory of the sample matrix. Therefore calibration can frequently be done without having multiple reference materials "strictly" matrix matched. Thus, GDMS is particularly suited for impurity survey of "4N to 6N" materials for which certified reference values are unavailable, like most of the elements in nuclear grade graphite. The latest generation of GDMS instruments (Model Element GD, Thermo Fisher Scientific) is equipped with a fast-flow ionization source (Figure 1) linked to a high-resolution sector-field mass analyzer. The FF-GDMS is a high power ionization source that greatly improves the atomization efficiency, often by one to two orders of magnitude in comparison to the previous generation of GDMS instruments. This makes a routine survey analysis significantly more efficient. In addition, the shortened analyses times allow evaluations of elemental distributions in relatively large sampling amounts/volumes with high sensitivity. Over the past decades, significant advancement in understanding the material behaviour in the GDMS environment, and thus in optimization of detection parameters has been achieved[15,16]. It is now possible to simultaneously quantify multiple trace elements in graphite in sensitive and rapid fashion.

Microwave induced combustion [17] is the most recent approach for combustion of samples. This closed vessel method is applicable also for completely digesting graphites including the nuclear grades. MIC is typically conducted in a pressurized thick-wall quartz vessel with oxygen gas (Figure 2). The sample/igniter mixture is placed on a quartz basket above the absorbing solution and is ignited by a high power microwave radiation. The analytes are then recovered in a refluxing cycle by the absorbing solution. This closed and low blank digestion procedure cleared a major hurdle on solution analysis of graphite that has haunted the analytic chemists for decades.

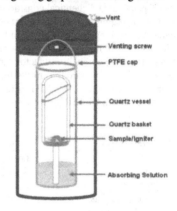

In this study, we launched a preliminary investigation on quantifying trace level contaminants in a nuclear-grade graphite sample by a direct and a wet sampling test method. Thus, selective trace elements, which are deemed important to the nuclear industry, were quantified by FF-GDMS and by MIC-ICP-MS. The limits of detection, uncertainty and certain limitations of the above methods were discussed.

Figure 2: MIC digestion apparatus

EXPERIMENTAL

FF-GDMS Analysis

Samples for this study (designated as NG-I) were prepared from a nuclear grade graphite rod (SGL Group). The analytical surface of all tested samples was first mechanically cleaned using a diamond pad, then further cleaned by designated Ta scribers and finally smoothed with clean filter papers (VWR, Inc.). Approximately 1-mm thick layer was removed from the cut surface on all analytical sides. The sample was cooled in the GD source by Peltier cooling then pre-sputter cleaned for 10 min prior to data acquisition, both under the same analytical conditions (Table II). The ion counting detector efficiencies were checked using well-characterized quality control samples. Data for the graphite sample were collected until the last three mass fraction readings varied by no more than 20%. Elements were scanned individually with an integration time of 10 – 100 ms. The average sampling amount is ~ 3 - 4 mg per measurement. Figure 3 shows a typical crater area formed on the graphite sample after plasma sputtering.

Figure 3: Typical plasma crater area on NG-I after 20 min analysis

Due to the wide dynamic range of the FF-GDMS instruments, the ion signals for all ionized sputtered species can be easily detected and consequently summed. Thus, if the effective ion beam intensity of the analyte ion (E) and the sum of the effective ion beam intensities for matrix ions detected (C), both corrected for isotopic abundance, are expressed in ratio, this ion-beam ratio (IBR) will be related to real mass fractions of atoms in the analyzed sample. This IBR of quantification by FF-GDMS methods would be an absolute (i.e. standard-less) method if there were no difference in ionization yields in the GD source among the elements. However, in a given matrix this difference is described by a proportionality factor, the so-called sensitivity factor or relative sensitivity factor (RSF) if related to that of the matrix element. Since the RSF varies from element to element, it has to be determined for accurate mass fraction measurements. The set of RSF values are determined by analyzing reference materials (RMs), $RSF(E) = Certified Mass Fraction of E in RM / IBR(E/C)$, by comparative analysis or by theoretical calculations from "*a priori*" physical and chemical considerations, if possible, determined by using RMs. All analytical instruments require this sort of calibration. However, the GDMS methods have the following advantages: a) empirically it has been found that the RSF values for most elements are close to unity and/or differ only over a narrow range; b) the RSF value for a given element is characteristic to that element, i.e. sensitivity factors are not influenced strongly by the character of the matrix; c) the calibration curves are linear over the entire mass fraction range. Consequently, if for an unknown analyte the RSF value is selected to be equal to one (1) in a particular matrix, the maximum bias in the IBR derived mass fraction results will be factor of two to three.

Table II. Instrumentation and analytic parameters for FF-GDMS measurement

Parameters	Values
Model	Thermo Element GD
GD Source:	Fast flow GD Assembly
Insulator:	Alumina
Discharge Gas:	Argon (6N+)
GD Conditions:	1.0 kV, 50 mA
Normalization:	Carbon as 100%
Detectors:	Faraday - 10 ms; Analog ~ 10 - 100 ms (isotope dependent)
Matrix Ion Current:	$> 1.0 \times 10^{-11}$A for Carbon
Mass Resolution:	~ 4000

In this work, all mass fraction results by FF-GDMS were evaluated using sensitivity values from the Carbon RSF set. This Carbon RSF set is based on the Standard RSF set, which is installed on all commercial GDMS instruments. The Standard RSF set is generally used as a "universal" set for routine / semi-quantitative analyses. The applicability of the sensitivity factors from the Standard RSF set for quantitative measurements of a particular matrix has to be verified and/or modified if required by analyzing certified materials or reference samples. The Carbon RSF set used in this study was derived after numerous measurements of solid carbon samples, graphites and metallic carbides previously analyzed by variety of test methods, such as INAA, XRF or ETV-ICP-MS.

MIC Digestion of Graphite and ICP-MS Measurement

Ultra-trace® - grade HNO_3 and NH_4NO_3 (VWR, Inc.) were used for MIC digestion. All vessels were thoroughly cleaned with soap solution, 10% HNO_3 and 18.2 MΩ H_2O.

On the basis of the procedure described by Flores, et al.[17], a modified MIC protocol was reported here for the complete digestion of nuclear grade graphite. The sample was prepared in the powder form by scrapping with a tantalum blade from a fresh surface of the graphite bar. For each digestion, ~50 mg graphite powder was weighed and transferred into a parafilm envelope (1×1 cm) that had been cleaned with 10% HNO_3 and 18.2 MΩ H_2O. Subsequently, 100 μL of 3 M NH_4NO_3

was added. After impregnation for ~ 1 h, the parafilm-wrapped sample was placed on a quartz basket (Scheme 2), which in turn was sealed in the quartz vessel containing 10 mL of HNO_3 (67- 70%) absorbing solution. The blank vessel was prepared in the similar manner except that no graphite was added. The quartz vessels were then mount on a microwave rotor (Multiwave 3000, Anton Paar) and pressurized to ~ 20 bar of oxygen (purity > 99.6%). The microwave radiation was applied at the maximum power (1400 W, 1.5 min) to trigger the combustion. Ignition typically occurred after ~1 min, evidenced by the rapid rising of pressure (up to 40 bar). Thereafter, a refluxing step (800 W, 90 min) was immediately followed to finish off the digestion. The residual carbon content is typically less than 1%.

ICP-MS measurement was performed on a PerkinElmer ELAN DRC II ICP-Mass Spectrometer with the following parameter settings: nebuliser gas flow 0.88 L/min argon, auxiliary gas flow 1.2 L/min argon, plasma gas flow 15 L/min argon, lens voltage 7.25 V, ICP RF power 1450 W. At least three standards were prepared and used to bracket the sample concentrations and all samples were acid matched. Calibration verification solutions were prepared and used as to fall in the middle of the calibration standard range.

RESULTS AND DISCUSSION

Elements like B, Mg, Al, Ca, Ti, V, Cr, Mn, Fe, Co, Ni, Cu, Zn, Zr, Mo, Sb, W and Pb in nuclear-grade graphite are of particular interest to the nuclear industry. Figure 4 summarizes the contents of these elements determined by FF-GDMS and MIC-ICP-MS methods. For comparison,

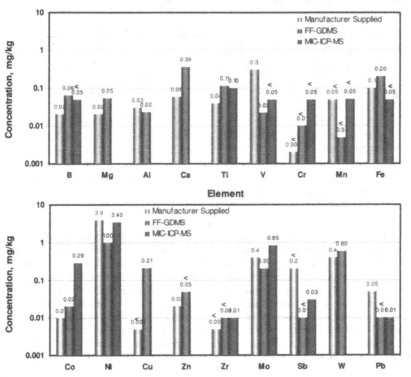

Figure 4: Trace element content of NG-I determined by FF-GDMS (average of 12 runs), MIC-ICP-MS (average of 2 runs) and reference value from the manufacturer.

the manufacturer reference values were also included. Both methods show narrow variations from the manufacturer's data on measuring B, Ti, Cr, Mn, Fe, Zr, Mo, Sb and Pb at sub-mg/kg levels. For example, all indicate that the B level below 0.1 mg/kg, Ti between ~ 0.04 and 0.11 mg/kg, and Pb between 0.05 and 0.01 mg/kg. For Mg, Al, Zn and W, FF-GDMS values were also consistent with the manufacturer reference values. MIC-ICP-MS analysis was not performed on these elements due to the high blank value. Source of interference is currently under investigation.

Discrepancy was observed on elements Ca, V, Co and Cu. The reference Ca value is ~0.06 mg/kg, and FF-GDMS found 0.36 mg/kg. For ICP-MS analysis, it was observed that significant amount of calcium was leached out from the quartz vessel during MIC digestion process, resulting up to 10 mg/kg Ca in the digestion solution. Therefore credible determination of Ca by MIC-ICP-MS could not be achieved. Both FF-GDMS and ICP-MS gave low V values (0.02 mg/kg and <0.05 mg/kg, respectively), in comparison to the reference ~0.3 mg/kg V. ICP-MS found Co at hundreds of μg/kg level, whereas FF-GDMS and the reference values are at tens of μg/kg level. With regard to Cu, FF-GDMS found ~ 200 μg/kg, which is much higher than the reference value <5 μg/kg. For Ni, MIC-ICP-MS value (~3.40 mg/kg) is higher than that obtained by FF-GDMS (1.0 mg/kg), but agrees well with manufacturer specification (~3.9 mg/kg). It should be noted that for FF-GDMS measurements, the instrumental parameters were set for routine multi-element analysis of graphite samples, rather than specifically tuned for determination of a few critical analytes. Therefore, uncertainty could arise for some elements. Table III summarizes the limit of detection (LOD) and uncertainty (i.e., SD) of FF-GDMS and ICP-MS measurements. Large uncertainties were noticed for ICP-MS measurement of Ni and Mo.

Table III. Limit of detection (mg/kg) and uncertainty

Element	FF-GDMS		MIC-ICP-MS		Element	FF-GDMS		MIC-ICP-MS	
	LOD	SD	LOD	SD		LOD	SD	LOD	SD
B	0.05	0.05	0.05	< 0.05	Co	0.005	0.004	0.05	0.02
Mg	0.05	0.012	-	-	Ni	0.005	0.47	0.05	0.35
Al	0.01	0.016	-	-	Cu	0.05	0.27	-	-
Ca	0.05	0.28	-	-	Zn	0.05	<0.01	-	-
Ti	0.005	0.114	0.05	0.14	Zr	0.01	<0.01	0.05	0.01
V	0.005	0.010	0.05	< 0.05	Mo	0.01	0.06	0.05	0.64
Cr	0.01	0.008	0.05	< 0.05	Sb	0.01	<0.01	0.05	0.02
Mn	0.005	0.001	0.05	< 0.05	W	0.01	4.74	-	-
Fe	0.005	0.15	0.05	< 0.05	Pb	0.01	0.001	0.05	0.01

Another factor causing the variation in element content determination between methods may be ascribed to the different sampling size or analyzed amount, due to the non-uniform distribution of contaminants. FF-GDMS samples 8 mm diameter spots with tens of microns in depth, whereas MIC-ICP-MS measures the average of orders of magnitude larger volumes. Figure 5 illustrates the distribution of tungsten and calcium from twelve sampling spots of the analyzed NG-1 graphite bar by FF-GDMS method. For instance, most measurements show sub-mg/kg level tungsten, but some unusually high values (e.g., 15 and 10 mg/kg W, etc) were also observed, indicating the presence of tungsten-rich regions or embedded particles. In comparison, most Ca readings are below 0.5 mg/kg, revealing that Ca element is rather uniformly distributed in this sample. The distribution of elements could be characteristic of the stages in manufacturing, or be altered by the mechanical sample preparation [14]. A detailed study of element distributions in this sample is currently undertaking.

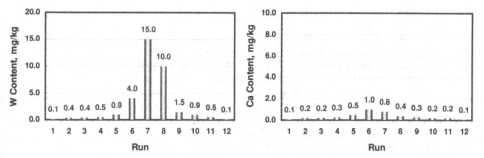

Figure 5: Distribution of Ca and W content in NG-I.

CONCLUSION

FF-GDMS (direct solid sampling) and MIC-ICP-MS ("wet" sampling) methods have been tested for determination of selective trace elements in nuclear grade graphite samples. Quantification of mg/kg – µg/kg level contaminants, including B, Mg, Al, Ca, Ti, V, Cr, Mn, Fe, Co, Ni, Cu, Zn, Zr, Mo, Sb, W and Pb, has been demonstrated on several test specimen. Both methods agreed well on the measurement of B, Ti, Cr, Mn, Zr, Sb and Pb, with values fluctuating mostly less than ±30%. In addition, these results indicate that these are fairly consistent with the "expected" total values from ash contents. It should be also noted here that the expected bias of GDMS results is generally within a factor of 2 to 3 of "true" values using the generalized Standard RSF / calibration sets for analyzing unknown materials. This distinguishing feature of the GDMS method is especially useful for survey analyses of materials for which there are no certified reference materials, such as nuclear grade graphites and/or purified manufactured carbons. This study also demonstrates that the FF-GDMS can be a powerful method for studying trace level distributions of elements in graphites and/or manufactured carbons.

Future efforts would be focused on identifying the source of interference, which currently limits the ability of MIC-ICP-MS on analyzing Mg, Al, Fe, Co, Zn, Mo and W, and on continuing optimization of the FF-GDMS parameters with materials cross-referenced by SIMS, INAA, ETV-ICP-OES and other methods, in particular for elements like Ca, Cu and Ni. Analysis on variety of nuclear-grade graphites is currently underway and results will be published elsewhere.

ACKNOWLEDGEMENT

We thank the EAG lead analyst Mr. Gabriel Infantino and Dr. Kwan H. Nam for their helpful discussion and assistance in ICP-MS measurements.

REFERENCES

(1) Bonal, J.-P.; Kohyama, A.; van der Laan, J.; Snead, L. L. In *MRS Bulletin*, 2009.
(2) Shibata, T.; Sumita, J.; Tada, T.; Hanawa, S.; Sawa, K.; Iyoku, T. *J. Nucl. Mat.* **2008**, *381*, 165.
(3) Virgilev, Y. S. *Atomic Energy* **1998**, *84*.
(4) Contescu, C.; Azad, S.; Miller, D.; Lance, M. J.; Baker, F. S.; Burchell, T. *J. Nucl. Mat.* **2008**, *381*, 15.
(5) Franzen, P.; Haasz, A. A.; Davis, J. W. *J. Nucl. Mat.* **1995**, *226*, 15.

(6)   Bolewski Jr., A.; Ciechanowski, M.; Dydejczyk, A.; Kreft, A. *Nucl. Instr. and Meth. in Phys. Res. B* **2005**, *237*, 602.
(7)   Delle, W.; Koizlik, K.; Nickel, H. *"Graphitische Werkstoffe fur den Enisatz in Kernreaktoren"*, Ed. Karl Thiemig A.G., pp 55-56, Munich, 1983.
(8)   Bogershausen, W.; Cicciarelli, R.; Gercken, B.; Konig, E.; Krivan, V.; Muller-Kafer, R.; Pavel, J.; Seltner, H.; Schelcher, J. *Fresenius J. Anal. Chem.* **1997**, *357*, 266.
(9)   Schaffer, U.; Krivan, V. *Anal. Chem.* **1999**, *71*, 849.
(10) Pickhardt, C.; Becker, J. S. *Fresenius J. Anal. Chem.* **2001**, *370*, 534.
(11) Schaffer, U.; Krivan, V. *Fresenius J. Anal. Chem.* **2001**, *371*, 859.
(12) Braun, T.; Rausch, H. *Anal. Chem.* **1995**, *67*, 1517.
(13) Mahantl, H. S.; Barnes, R. M. *Anal. Chem.* **1983**, *55*, 403.
(14) Michellon, C.; Putyera, K.; Kasik, M.; Hockett, R. *"Production Support and Process Control of PV Materials by Direct Sampling High Resolution Glow Discharge Mass Spectrometry Methods"*, Application Notes PA117, Evans Analytic Group LLC, 2008.
(15) Putyera, K.; Su, K.; Liu, C.; Hockett, R. S.; Wang, L. 2008 Fall MRS Proceedings - Photovoltaic Materials and Manufacturing Issues, Vol.1123, Paper#1123-P01-08, Dec 1-5 2008, Boston MA.
(16) Spitsberg, I. T.; Putyera, K. *Surf. & Coatings Techn.* **2001**, *139*, 35.
(17) Flores, E. M. M.; Barin, J. S.; Mesko, M. F.; Knapp, G. *Spectrochimica Acta Part B* **2007**, *62*, 1051.

Mater. Res. Soc. Symp. Proc. Vol. 1215 © 2010 Materials Research Society   1215-V16-15

Thermal Stability of Microstructure in Grain Boundary Character Distribution-Optimized and Cold-Worked Austenitic Stainless Steel Developed for Nuclear Reactor Application

Shinichiro Yamashita[1], Yasuhide Yano[1], Ryusuke Tanikawa[2], Norihito Sakaguchi[2], Seiichi Watanabe[2], Masanori Miyagi[3], Shinya Sato[3], Hiroyuki Kokawa[3]
[1] Oarai Research and Development Center, Japan Atomic Energy Agency (JAEA), 4002, Narita-cho, Oarai-machi, Ibaraki, 311-1393, Japan
[2] Center for Advanced Research of Energy Conversion Materials, Hokkaido University, N-13, W-8, Kita-ku, Sapporo 060-8628, Japan
[3] Department of Materials Processing, Graduate School of Engineering, Tohoku University, 6-6-02 Aramaki-aza-Aoba, Aoba-ku, Sendai, 980-8579, Japan

ABSTRACT

Grain boundary character distribution-optimized (GBCD) Type 316 corresponding austenitic stainless steel and its cold-worked form (GBCD+CW) are prospective materials to be considered for next generation nuclear energy systems. Specimens of these steels were thermally-aged at 973 K for 1 and 100 h and then examined by transmission electron microscopy (TEM) to evaluate microstructural stability during heat treatment high temperature. TEM results revealed that microstructures of both specimens types prior to ageing contained step-wise boundaries which were composed of coincidence site lattice (CSL) boundaries. The GBCD+CW specimens had dislocation cells and networks as well as deformation twins whereas the GBCD one possessed few dislocations. After thermal ageing, the precipitates were formed on not only random grain boundaries but also on dislocations, and they contribute to prevent significant microstructural change such as recrystallization and dislocation recovery.

INTRODUCTION

Austenitic stainless steels have been commonly used as structural materials in power-generating industries due to their high ductility and fracture toughness, but meanwhile successive efforts to improve their properties have also been made to meet the severe requirement for structural materials used in future nuclear reactors, such as advanced light water reactors and fast breeder reactors [1,2].

Among several approaches to improve the material property, grain boundary structure studies using commercial austenitic stainless steels have suggested prospective and practical methods to modify the bulk property of the steels; for example, the optimal distribution of coincidence site lattice (CSL) boundaries and consequent discontinuity of random boundary network was demonstrated to enhance intergranular corrosion resistance [3-6]. Of particular interest is that this property is relatively easy to obtain through simple thermo-mechanical treatment process without any change of the original chemical composition.

In this study, the thermal stability of microstructures of the grain boundary character distribution-optimized (GBCD) Type 316 austenitic stainless steel and its cold-worked corresponding form (GBCD+CW) was investigated, mainly focusing on the effects of the additional cold-working and precipitate formation on the microstructural stability during high temperature thermal treatments.

EXPERIMENT

The material used was Ti-modified Type 316 austenitic stainless steel which was originally developed for fast reactor core application; a typical composition would be Fe-16Cr-14Ni-0.05C-2.5Mo-0.7Si-0.025P-0.004B-0.1Ti-0.1Nb in wt% [7,8]. This material, solution heat-treated at 1373 K for 0.5 h, was termed the starting base material (BM) in this study. A BM specimen, 25 x 12 x 1.7 mm^3 in size, was thermo-mechanically processed by 3% cold-rolling and subsequent annealing at 1380-1420 K for 3 h to give a GBCD specimen. Details of the thermo-mechanical processing conditions can be found elsewhere [9,10]. Following the thermo-mechanical treatment, further additional cold-rolling ranging from 4 to 20% was conducted on individual GBCD specimens to obtain the cold-worked (GBCD+CW) specimens.

The frequency of CSL boundaries and grain boundary character distributions were assessed by orientation imaging microscopy (OIM). In this study, grain boundaries with $\Sigma \leq 29$ were regarded as low-Σ CSL boundaries, and Brandon's criterion [11] was adopted for the critical deviation in the grain boundary characterization [12].

Small coupons made from the GBCD and the GBCD+CW specimens were thermally-aged at 973 K for 1 and 100 h. And then microstructures prior to and post ageing were characterized by transmission electron microscopy (TEM) observations using the JEOL 4000FX microscope operated at 400 kV.

RESULTS AND DISCUSSION

Grain boundary character distribution

The grain boundary character distributions of a BM specimen and of specimens thermo-mechanically processed by 3% cold-rolling and subsequently annealed at 1380-1420 K for 3 h are presented in figure 1. The frequency of CSL boundaries in figure 1(b-d) of the GBCD specimens thermo-mechanically processed reached 81.3-82.9%, whereas the frequency in the BM specimen (figure 1(a)) was 45.5%. The network of random grain boundaries seen in the BM

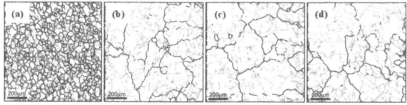

Figure 1. Grain boundary character distributions shown by OIM images for (a) BM specimen and (b-d) thermo-mechanically processed Ti-modified Type 316 austenitic stainless steel specimens with 3% cold-rolling and subsequent annealing for 3 h at (b) 1380, (c) 1400, and (d) 1420 K. Random and CSL boundaries are indicated by thick black and thin gray lines, respectively.

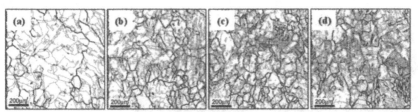

Figure 2. Grain boundary character distributions shown by OIM images for GBCD+CW specimens. The extents of additional cold-rolling were (a) 7, (b) 10, (c) 16, and (d) 20%.

specimen was totally disrupted in all of the GBCD specimens and short random boundary segments were isolated. The grain boundary character distributions of the GBCD+CW are shown in figure 2. The ratio of gray regions in the maps increased with increasing amounts of cold rolling. More than 70% of the CLS boundaries were retained even after the additional cold-rolling.

Microstructures prior to and post thermal ageing

Typical microstructures of (a) GBCD, (b) GBCD+CW7% and (c) GBCD+CW20% prior to thermal ageing are shown in figure 3. All microstructures were regarded as containing step-wise boundaries, all of which were composed of a combination of CSL boundaries. In addition, grain boundaries including step-wise ones seemed to be sharp and clear. No precipitate formation was found on any boundaries. The difference in the cold-rolling ratio appeared because of dislocation cells, network dislocation and deformation twins, all of which seemed to be related to the amount of strain introduced into the matrix.

Representative thermally aged microstructures of GBCD and the GBCD+CW specimens are shown in figure 4. After ageing at 973 K for 100 h, precipitates formed on grain boundaries. Also, more precipitates formed on high energy random boundaries than on low energy CSL boundaries. Sometimes, they preferentially formed on relatively high energy CSL boundaries with lower regularity among the CSL ones as it is seen in step-wise boundaries of figure 4 (c).

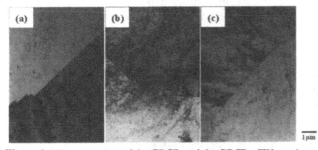

Figure 3. Microstructures of the GBCD and the GBCD+CW specimens; (a) GBCD, (b) GBCD + CW7%, and (c) GBCD+ CW20%.

123

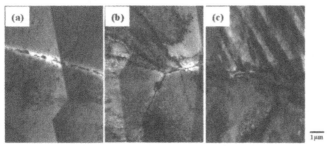

Figure 4. Microstructures after thermal ageing at 973 K for 100 h of (a) GBCD, (b) GBCD+CW7%, and (c) GBCD+CW20% specimens.

It should be noted here that matrix defect structures such as dislocation cells, network dislocation and deformation twins remained after the intergranular precipitate formation caused by annealing at high temperature. This fact implied that the effect of cold-rolling on radiation defects would be highly expected for GBCD+CW specimens with a high frequency of CSL boundaries because radiation defect sinks such as dislocation cells, network dislocation and deformation twins introduced by cold-rolling were stabilized due to the intergranular precipitate formation.

Precipitate formation

Figure 5 shows microstructures of precipitates formed on step-wise boundaries after ageing at 973 K for 100 h in GBCD and GBCD+CW7% specimens. The characteristic of step-wise boundaries are clearly visible; the continuity of intergranular M_6C-type carbide precipitates, which tend to form preferentially on less ordered or random boundaries [13], was well fragmented by CSL boundaries. On the other hand, precipitate formation within the grain was also observed after thermal treatment at high temperature.

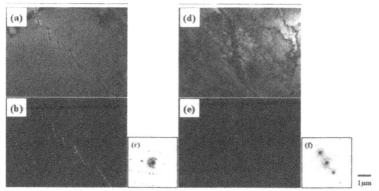

Figure 5. Precipitates formed on step-wise boundaries after ageing at 973 K for 100 h in (a-c) GBCD and (d-f) GBCD+CW7% specimens. (a, d) are bright field images; (b, e) are dark field images; (c, f) are electron diffraction patterns.

124

Figure 6 shows fine precipitates formed on a dislocation network after ageing at 973 K for 100 h in GBCD+CW7% and GBCD+CW20% specimens. From diffraction pattern analyses, both precipitates were identified as the same titanium-rich MC carbide. Previous research and development experiences in structural materials applicable to nuclear reactors [14] have shown that the titanium-rich MC carbides play a very significant role in stabilizing the dislocation structure and they has been commonly used as an effective method for improving the swelling resistance property in this class of steel. Therefore, even in the GBCD+CW specimens, it could be highly anticipated a similar effect of the MC carbide on the dislocation structure.

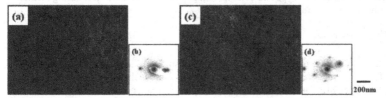

Figure 6. Precipitates formed on a dislocation network after ageing at 973 K for 100 h for (a, b) GBCD+CW7% and (c, d) GBCD+CW20%; (a, c) are dark field images; and (b, d) are electron diffraction patterns.

Prospective effect of GBCD optimization on the material property

Microstructural evolutions are schematically illustrated in figure 7, based on the OIM and TEM results for both the GBCD and GBCD+CW specimens during thermal treatment at high temperature. The microstructure of both specimens indicated high thermal stability. In addition, it was possible to expect the following two potential effects for applications as nuclear reactor structural materials. The first effect is a good creep resistance at elevated temperature compared with the existing non-GBCD materials because of the higher resistance for grain boundary sliding attributed to random boundary network fragmentation. The second effect is an improved radiation tolerance due to the increment of the density of radiation defect sinks that accompanied the intragranular MC-type carbide precipitates, which stabilized the dislocation structure. Further studies are required to investigate the GBCD optimization effect on the long-term high temperature mechanical property and on the dimensional stability during high energy neutron irradiation to high dose.

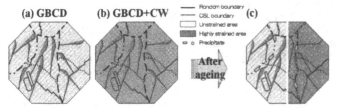

Figure 7. Schematic illustration of the microstructures of the GBCD and the GBCD+CW specimens before and after ageing. (a) and (b) show the microstructures of the GBCD and the GBCD+CW specimens before ageing and (c) shows both after ageing.

CONCLUSION

Thermal annealing experiments and subsequent microstructural observation showed that both GBCD and the GBCD+CW specimens retained high thermal microstructural stability due to a thermo-mechanical process, within the testing conditions of this study. Precipitates identified after a thermal ageing experiment at 973 K were M_6C type carbides on segmented random boundaries and titanium-rich MC-type carbides on dislocation structures (network dislocations and dislocation cells); These precipitates were responsible for the stabilization of the microstructures of the GBCD Type 316 corresponding austenitic stainless steel during thermal treatment at high temperature.

ACKNOWLEDGMENTS

The authors wish to thank Emeritus Professor Heishichiro Takahashi of Hokkaido University for his useful comments and Mr. Manabu Sekine of Nuclear Technology & Engineering Company for his assistance in TEM sample preparation and observation.

REFERENCES

1. American Society of Materials. ASTM handbook: fatigue and fracture. USA: Materials Information Society International; 1996.
2. P. Yvon, F. Carré, J. Nucl. Mater., 385, 217 (2009).
3. H. Kokawa, M. Shimada, and Y.S. Sato, JOM, 52 (7), 34 (2000).
4. M. Shimada, H. Kokawa, Z.J. Wang, Y.S. Sato, I. Karibe, *Acta. Materialia.* 50, 2331 (2002).
5. M. Michiuchi, K. Kokawa, Z.J. Wang, Y.S. Sato, K. Sakai, *Acta. Materialia.* 54, 5179 (2006).
6. H. Kokawa, M. Shimada, M. Michiuchi, Z.J. Wang, Y.S. Sato, *Acta. Materialia.* 55, 5401 (2007).
7. T. Itaki, S. Yuhara, I. Shibahara, H. Kubota, M. Itoh, and S. Nomura, *Materials for Nuclear Reactor Core Applications* (BNES, London, 1987), p. 203.
8. I. Shibahara, S. Ukai, S. Onose and S. Shikakura, *J.Nucl.Mater.* 204. 131 (1993).
9. S. Yamashita, Y. Yano, S. Watanabe, N. Sakaguchi, R. Tanikawa, H. Kokawa, M. Miyagi, S. Sato, J. P. Patent No. 2008-024422 (4 February 2008)
10. S. Yamashita, Y. Yano, M. Endo, N. Sakaguchi, S. Watanabe, M. Miyagi, S. Sato, Y.S. Sato, H. Kokawa, M. Kawai, presented at the 2008 AESJ Fall Meeting, Kochi, 2008 (unpublished).
11. D. G. Brandon, *Acta Metallurgica*, 14, (11), 1479 (1966).
12. H. Kokawa, T. Watanabe, S. Karashima, Script Metall, 17, 1155 (1983).
13. E. A. Trillo and L. E. Murr, J. Mater. Sci., 33, 1263 (1998); Acta Materiailia, 47, 235 (1999).
14. Eal H. Lee, P. J. Maziasz, and A. F. Rowcliffe, *Phase Stability During Irradiation*, Conference Proceedings, edited by J. R. Holland, L. K. Mansur and D. I. Potter, (The Metallurgical Society of AIME, 1980), p.191.

Mater. Res. Soc. Symp. Proc. Vol. 1215 © 2010 Materials Research Society   1215-V16-18

Reprocessing Silicon Carbide Inert Matrix Fuel by Molten Salt Corrosion

Ting Cheng, Ronald H. Baney[1], James Tulenko[2]
[1]Material Science and Engineering Department, University of Florida, 32611, Gainesville, FL, U.S.A.
[2]Nuclear Engineering Department, University of Florida, 32611, Gainesville, FL, U.S.A.

ABSTRACT

Silicon carbide is one of the prime matrix material candidates for inert matrix fuels (IMF) which are being designed to reduce plutonium and long half-life actinide inventories through transmutation. Since complete transmutation is impractical in a single in-core run, reprocessing the inert matrix fuels becomes necessary. The current reprocessing techniques of many inert matrix materials involve dissolution of spent fuels in acidic aqueous solutions. SiC cannot be dissolved by that process. Thus, new reprocessing techniques are required.

This paper discusses a possible way for separating transuranic (actinide) species from a bulk silicon carbide (SiC) matrix utilizing molten carbonates. Bulk reaction-bonded SiC and SiC powder (1 μm) were corroded at high temperatures (above 850 °C) in molten carbonates (K_2CO_3 and Na_2CO_3) in an air atmosphere to form water soluble silicates. Separation of Ceria (used as a surrogate for the plutonium fissile fuel) was achieved by dissolving the silicates in boiling water and leaving behind the solid ceria (CeO_2).

INTRODUCTION

During the past decade, considerable efforts have been made to search for a material to be utilized as the matrix for inert matrix fuels (IMF) to transmute plutonium and minor actinides (Np, Am, Cm) in a nuclear reactor. The properties of silicon carbide attracted extensive attention to function as an IMF's due to the small thermal neutron absorption cross section of both silicon and carbon, its chemical inertness, and high thermal conductivity [1-4].

Issues related to the application of SiC for IMF's including fabrication, in-reactor behavior, reprocessing and waste disposal remained. Recently, S. Bourg [4] developed a separation method for SiC matrix fuel using Cl_2 to oxidize SiC powder (45 μm), in which 75 percent of SiC was volatilized in the form of $SiCl_4$ at 900 °C. The remaining solid carbon was removed by oxidation under oxygen at 400 °C. This paper did not consider the hazardous and corrosive nature of this process, if the system were applied on a large scale. In our work, separation can be achieved by filtering ceria, a widely used surrogate of plutonium [5-7], from a solution of water-soluble SiC corrosion products (e.g. silicate) formed by SiC corrosion in molten carbonates.

Experimental

In our experiment, a tube furnace with caps on both ends was preheated to 900 °C for SiC powder and 1050 °C for SiC monoliths contained in carbonate baths. In all experiments, a compressed air cylinder provided constant airflow (658 ml/min) through the furnace. SiC powder (purchased from Alfa Aesar) with 1 μm diameter was studied. After ball milling for 9 hours, the mixture of SiC powder (1 g) and Na_2CO_3 at approximately a 1:2 molar ratio was distributed in three alumina crucibles which were placed midway in the tube furnace. Powder samples were held for various times up to 0.5 hours, after which they were rapidly quenched by immersing them into boiling de-ionized water to dissolve corrosion products and residual Na_2CO_3. The solution and residue were transferred to centrifuge tubes, centrifuged, dried and the unreacted SiC was weighed. The average SiC weight loss percent was determined.

Reaction bonded SiC 5.0 mm diameter rods (purchased from Goodfellow Corporation; theoretical density: 95.1%) were cut into three pellets with a uniform thickness of 2 mm. The pellets were then ground with fine grit grinding paper to remove any tool markings. The pellets were corroded in molten K_2CO_3 (molar ratio of salt to SiC is 10). Weight loss of the pellets was measured periodically after washing in boiling water and drying. The residue of the pellets was put back into the furnace with fresh salt with the same molar ratio to SiC.

Experiments on ceria powder (5-9 μm purchased from Alfa Aesar) corroded by alkali salts (K_2CO_3 and Na_2CO_3) at a 1:10 molar ratio (CeO_2: salt) were performed at 1050 °C for 40 hours. Three parallel tests were conducted for each reactant combination to obtain the average value. The weight difference was measured between the original ceria and the residue ceria after corrosion and washing away the alkali salt. Blank experiments were also performed, in which the ceria weight change resulted from washing and transferring the powder from one container to another was calculated to determine error of the weight change measurement.

DISCUSSION

SiC under oxidation is covered by a thin silica layer which protects it from further oxidation. When exposed to molten carbonates, silica is easily reacted to form crystalline or molten silicates. Further oxidation of the SiC surface occurs as oxygen diffuses through the silicate layer.

Complete corrosion of 1 μm SiC powder was achieved in 30 min. The average weight loss percentage of SiC powder (1 μm) as a function of reaction time is shown in figure 1.

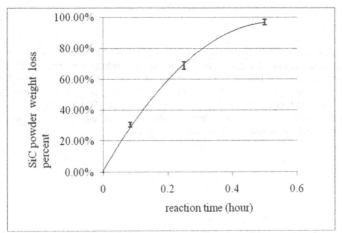

Figure 1 SiC powder (1 μm) weight loss percent as a function of the reaction time at 900 °C

Operation error due to transferring and washing was negligible. Sodium silicate (Na_2SiO_3) was identified by XRD to be the major product. See figure 2.

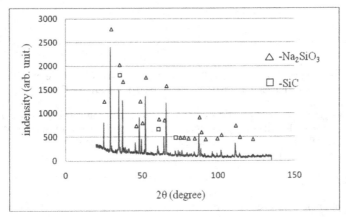

Figure 2 X-ray diffraction spectrum of product of SiC powder corrosion in Na_2CO_3 molten salt at 900 °C for 30 minutes

Experiments to dissolve bulk reaction-bonded (RB) SiC pellets were conducted, since grinding spent bulk SiC into powder is not practical in nuclear industry due to radiological contamination of grinding equipment and loss of powder to the atmosphere. No weight loss of RB-SiC in Na_2CO_3 was observed at 900 °C. Several papers [8-12] reported that once the silicate layer grows to a critical thickness and the temperature is sufficiently low, molten salt ions may not pass through the thick silicate layer. As a consequence, the corrosion would be terminated. If the temperature is considerably higher, i.e. above the melting temperatures of the silicates as determined in published literate [9], the alkali silicate layer will be dissolved in molten carbonates. Ions would diffuse faster in the melts than in the solids, hence corrosion rate would be enhanced.

Corrosion temperature for bulk SiC was increased to 1050 °C in order to melt the silicate layer formed on SiC surface. Complete dissolution of SiC pellets, as seen in figure 3, proved the feasibility of utilizing a molten sat corrosion method.

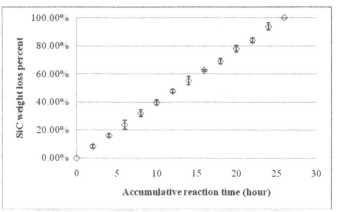

Figure 3 Weight loss percent of SiC pellets as a function of accumulative reaction time in K_2CO_3 molten salt at 1050 °C (Molar ratio of K_2CO_3 to SiC=10)

Weight loss of ceria powder after corrosion in molten salt is negligible and approximate to the ceria weight loss caused by operation error. This result confirms that ceria does not react with molten salt at 1050 °C. It is predicated that plutonium is unsusceptible to K_2CO_3 melt at 1050 °C due to its similar properties to ceria.

CONCLUSIONS

Application of SiC as an inert matrix for nuclear fuel is hindered because no efficient method for separating the unspent fuel from SiC matrix has been indentified to date. A feasible method in which the SiC matrix was dissolved in molten salt while leaving the surrogate unreacted has been potentially developed. It is complicated to quantitatively estimate the time for complete corrosion of certain amount of bulk SiC, since corrosion rate may be affected by multiple parameters such as surface to volume ratio, fabrication method and density of SiC samples, reaction atmosphere, and temperature and so on. Investigation of those factors is ongoing. In conclusion, more research needs to be carried out to determine the optimal utilization of the temperature and proportions and utilization of air with the K_2CO_3 dissolution material. However, initial research indicates that the combination of materials holds promise to offering a solution to the reprocessing of spent SiC matrix fuel for nuclear reactors.

ACKNOWLEDGEMENTS

The authors gratefully thank DOE to financially support this project (E-FC07-06ID14741).

REFERENCES

1. C. Degueldre, J.M. Paratte, J. Nucl. Mater. 274 (1999) 1-6.
2. R. Naslain, Comp. Sci. Tech. 64 (2004) 155-170.
3. H. Matzke, V.V. Rondinella and T. Wiss, J. Nucl Mater. 274 (1999) 47-53.
4. S. Bourg, F. Peron and J. Lacquement, J. Nucl Mater. 360 (2007) 58-63.
5. U. B. Pall, C. J. MacDonald, E. Chiang, W. C. Chernicoff, K. C. Chou, M. A. Molecke, D. Melgaard, Metallug Mater. Trans. 32 (2001) 1119-1128.
6. Y. S. Park, H. Y. Sohnb and D. P. Butt, J. Nucl. Mater. 280 (2000) 285-294.
7. A. Tatsumi, I. Kazuya, Y. Kazuhiro, T. Satoshi, I. Yaohiro, J. Alloys Compd. 394 (2005) 271-276.
8. M. S. Jacobson, J. Amer. Ceram. Soc. 69 (1986) 74-82.
9. M. D. Allendorf and K. E. Spear, J. Electrochem. Soc. 148 (2001) B59-B67.
10. S. Y. Lee, Y. S. Park, P. Hsu, and M. J. McNallan, J. Amer. Ceram. Soc. 86 (2003) 1292-1298.
11. A. Kosminski, D.P. Ross and J.B. Agnew, Fuel Proc. Tech. 87 (2006) 1037-49.
12. A. R. Jones, R. Winter, G. N. Greaves and I. H. Smith, J. Phys. Chem. 109 (2005) 23154-61.
13. D. P. Dobson, A. P. Jones, R. Rabe, T. Sekine, K. Kurita, T. Taniguchi, T. Kondo, T. Kato, O. Shimomura, S. Urakawa, Earth Planet. Sci. Lett. 143 (1996) 207-215.
14. M. Mohamedi, Y. Hisamitsu, I. Uchida, J. Appl. Electrochem. 32 (2002) 111-117.
15. V. Volkovich, T. R. Griffiths, P. D. Wilson, J. Chem. Soc. 92, (1996) 5059-5065.

Mater. Res. Soc. Symp. Proc. Vol. 1215 © 2010 Materials Research Society   1215-V16-22

Evaluation of Adhesive Strength of Oxide Layer on Carbon Steel at Elevated Temperatures

Manabu Satou
Department of Quantum Science and Energy Engineering, Tohoku University,
6-6-01-2, Aramaki-aza-Aoba, Sendai, 980-8579, JAPAN

ABSTRACT

Adhesive strength of the oxidation layers on carbon steel was evaluated by means of a laser shock method, which uses a pulsed laser to generate shock wave. Oxidation for 200 hours in air created 10-micron-thick magnetite on carbon steel. Typical strength of the layer was evaluated to be about 50MPa at ambient temperature. The adhesive strength was varied from around one-tenth of yield stress to the ultimate tensile strength of the base materials. The adhesive strength of the oxide layer depended on test temperature. It is possible that the adhesive strength becomes an essential parameter for the evaluation of the protective layers.

INTRODUCTION

Adhesive strength of the layers that protect from corrosive environments for structure materials in nuclear systems including fusion and fission reactors is of interest not only from fundamental point of view but also engineering point of view, because the protective function of the layer is effective when the layers maintain on base materials with suitable strength. For example, although lead-bismuth eutectic coolant in fast reactor design provides superior advantages from view points of neutronics and safety issue, severe corrosion of structure material is one of the key issues to be solved[1]. Formation of the protective oxide layer depends on oxygen potential in lead-bismuth as the coolant material and on chemical composition of steel as the structure material. It has been demonstrated that the contents of chromium in the steel promote the formation of stable oxide layer. Once oxide layers were formed on the surface in the coolant material, corrosion behavior changed from the liquid metal corrosion to the oxidation corrosion [2,3]. Formation of stable chromium-rich oxide layer is believed to prevent break away of the oxide layer. Stability of the oxide layer from a view point of the thermodynamics is important as well as the mechanical properties of the oxide layer including adhesive properties against the base material. In this paper, an attempt was carried out to evaluate the adhesive property of the protective coating layer on the structure materials, as for a simple example, oxidation layer on a carbon steel was examined.

EXPERIMENTAL

Evaluation method of the interface strength depends on degree of the applied stress and its mode such as shear, tensile or compression. Environment like temperature and atmosphere will affect the evaluation method. Frequency of the stress and temperature cycles and duration of the applying stress will affect its fracture modes such as impact fracture or creep failure. Integrity of the interface depends on manufacture procedures. Resultant secondary stress due to differences of the thermal expansion also affects the interface properties. In this paper, to simplify the applied stress condition, evaluation was carried out at tensile stress mode at relatively fast strain rate for the oxide film obtained on the carbon steel at an elevated temperature in air. The fast strain rate stress condition should contribute to preventing additional deformation near the interface of the layer and the base metal.

Sample preparation and characterization

Disk shaped specimens with diameter of 20mm and thickness of 2mm were prepared to measure the adhesive strength between the oxide layer and the carbon steel. Oxidation layer on the specimen made of the carbon steel JIS (Japanese Industrial Standards) G4051 S45C "Carbon steels for machine structure" was obtained by heat treatment at an elevated temperature in air for 10 to 200h. Nominal chemical composition of S45C is in table 1. Oxidation layer characteristics were examined by X-ray diffraction with Cu-K at 40mA and 40kV.

Table 1 Chemical composition of carbon steel S45C (nominal, in wt%).

C	Si	Mn	P	S	Fe
0.42-0.48	0.15-0.35	0.30-0.90	<0.030	<0.035	Bal.

Laser shock method to measure adhesive strength at interfaces

An experimental measurement technique of adhesive strength at interfaces utilized a pulse laser shock described elsewhere [4]. In this measurement, the compression shockwave caused by the laser pulse shot travels from back side of the oxide layer to the front side will reflect as tensile stress wave. When the tensile stress reached the maximum of the adhesive strength resulted in exfoliation the oxide layer. The corresponding stress defined as the adhesive strength of the layer. The apparatus used in this work contained two parts, a Nd:YAG laser with wave length of 1064nm and the maximum 1.6J power at single-pulse mode to apply tensile stress at the interface and a displacement laser interferometer using a single mode diode-pumped solid-state (DPSS) laser with wave length of 532nm as shown in fig.1. The first step is to determine the critical laser power induced the exfoliation of the oxide layer. A series of short pulsed laser irradiation was carried out over 3 mm-diameter in area on 60μm-thick water glass film that was the back surface of the substrate disk. The laser-induced expansion of the S45C under confinement by the water-glass generated compressive stress wave directed toward the oxide layer on the substrate's front surface. The compression stress wave reflected into tensile wave from the free surface of the oxide layer and lead to its exfoliation at sufficiently high amplitude of energy. The second step is to measure the free surface velocity during interface exfoliation at ambient temperature. The displacement interferometer showed that about 10 displacement fringes were produced during the wave propagation. Digital oscilloscope was used to record such fringes with a resolution of 0.05 ns. Typical PIN photodiode voltage output corresponding to the mirror finished S45C free surface obtained by the displacement interferometer is shown in fig. 2. The photodiode output voltage amplitude $A(t)$ is related to the fringe count $f(t)$ as:

$$A(t) = \frac{(A_{max} + A_{min})}{2} + \frac{(A_{max} - A_{min})}{2} \sin[2\pi f(t) + \delta], \qquad (1)$$

where A_{max}, and A_{min} are the maximum and the minimum fringe amplitudes, respectively, and δ is the phase angle. Once $f(t)$ is determined from the above, the displacement $u(t)$ can be obtained as:

$$u(t) = \lambda_0 f(t)/2, \qquad (2)$$

where λ_0 is the wavelength (532 nm) of the DPSS laser.

$$V(t) = \frac{du(t)}{dt} = C(e^{-t/A} - e^{-t/B}). \qquad (3)$$

Assuming the velocity $V(t)$ can be written as eq.(3), the three parameters are fitted to the measured output $A(t)$ as shown in fig.2. The typical velocity obtained from the fitting is shown in fig.3 depended on the laser power. The third step is calculation of the stress. For the oxide of density ρ and thickness h, the interface stress σ is calculated using the measured free surface velocity $V(t)$ as:

$$\sigma_f(x,t) = \rho c_f^2 \frac{\partial u}{\partial x} = \rho c_f \left\{ V(t + \frac{x}{c_f}) - V(t - \frac{x}{c_f}) \right\}, \qquad (4)$$

where, c_f is the longitudinal stress wave velocity in the oxidation layer. The velocity can be written using Lame's constants λ and μ as follow:

$$c_f = \sqrt{\frac{\lambda + 2\mu}{\rho}}. \qquad (5)$$

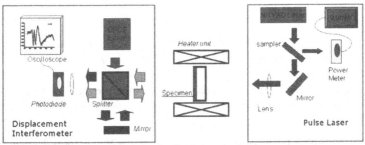

Fig. 1.   Schematics of laser shock adhesive strength measurement apparatus.

RESULTS AND DISCUSSION

Characteristics of oxidation layers

Fig.4. shows X-ray diffraction pattern from the specimens exposed in air at 500°C for 194h. Magnetite (Fe_3O_4) and hematite (Fe_2O_3) are identified. Typical thickness was measured from cross-sectional observation by an optical microscope. The thickness was well fitted to oxidation rate equation with parabolic law. About 10μm of thickness was obtained after about 200 hours at 500°C.

Adhesive property at ambient temperature

From a series of laser shots with various levels of laser energy, critical energy was

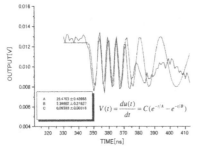

$$V(t) = \frac{du(t)}{dt} = C(e^{-t/A} - e^{-t/B})$$

Fig. 2.   Typical output from PIN photodiode for interferometer to measure surface velocity. (Laser power:1360mJ, Substrate: carbon steel)

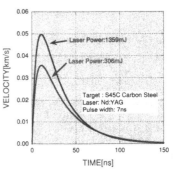

Fig. 3.   Surface velocity profile of carbon steel by laser shots with different power.

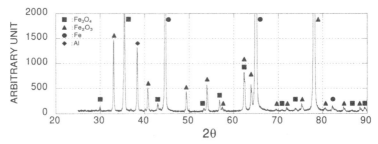

Fig. 4.   X-ray diffraction pattern of oxidation layer on carbon steel after heat treatment in air for 194 hours at 500°C. Square: Magnetite (Fe₃O₄), Triangle: Hematite (Fe₂O₃), Circle: Iron (Fe).

determined about 290mJ. Exfoliation was clearly observed at a 383mJ-shot as shown in fig. 5. The evaluated value of the adhesive strength was about 50MPa for the specimen prepared in air at 500°C for 194h. Elastic parameters of magnetite using the stress evaluation in eq. (4) were 230GPa for Young's modulus and 0.26 for Poisson's ratio. The typical yield stress or ultimate tensile strength of S45C is about 400 to 600MPa. Therefore adhesive strength of the oxide on S45C is about one-tenth of the strength. It is suggested that oxidation layer may exfoliated well before yielding.

Adhesive property at elevated temperature in lead-bismuth

Another series of laser shots exposed to lead-bismuth was carried out at 270°C. Disk specimens were faced to the melted lead-bismuth eutectic alloy in furnace and the laser shots were from the other side. The pictures in fig. 6 show surfaces of the S45C exposed to the lead-bismuth after laser shots at different energy. Small exfoliation indicated by white arrow is observed at laser shot of 1660mJ power. Exfoliation was hardly observed for the area irradiated

136

at 1340mJ. The critical level of energy for the exfoliation observed is rather high compared to ambient shots. The higher level of the critical energy indicates that the lead-bismuth might reduce the tensile stress applied to the interface of the oxide and S45C. Fig.7 shows X-ray diffraction pattern obtained after exposure to lead-bismuth at 270°C. Oxide formation of lead-bismuth is also observed. Surface morphology showed in fig.8 indicated dissolution of oxide might happen because the needle-like oxide became thinner as shown in the embedded picture. The tensile stress applied by the laser shot to thinner oxide layer resulted in less stress compared to thicker layer in this case [5]. Adhesive property at elevated temperature without lead-bismuth to reduce dissolution of the oxidation layer or protective layer prepared by controlled atmosphere and base material probably are needed for further consideration.

CONCLUSION

Adhesive property of the oxidation layer on the carbon steel was examined by means of laser shock method. Typical strength was determined to be 50MPa at ambient environment. An attempt was carried out at the elevated temperature exposed to the lead-bismuth eutectic alloy. Dissolution of the oxidation layer during the exposure may affect the evaluation of the adhesive property of the oxidation layer prepared in air at elevated temperature.

Fig. 6. Exfoliation of oxidation layer on carbon steel in lead-bismuth at 270°C. Exfoliation observed in (a) indicated by white arrow with laser power of 1660mJ, not (b) with laser power of 1340mJ.

Fig. 5. Exfoliation of oxidation layer on carbon steel by laser shot of 383mJ at ambient temperature.

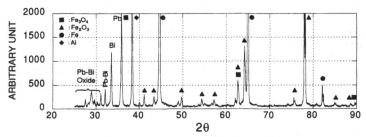

Fig. 7. X-ray diffraction pattern of oxidation layer on carbon steel after heat treatment in air for 194 hours at 500°C followed by exposure to lead-bismuth at 270°C. Square: Magnetite (Fe_3O_4), Triangle: Hematite (Fe_2O_3), Circle: Iron (Fe).

Fig. 8.   Surface morphology of oxidation layer on carbon steel before (a) and after (b)
exposure to lead-bismuth at 270°C

ACKNOWLEDGMENTS

Present study includes the result of "Development of evaluation method of the adhesive property of corrosion-resistant layer on structural materials in lead-bismuth cooled fast reactor" entrusted to Tohoku University by the Ministry of Education, Culture, Sports, Science and Technology of Japan (MEXT). This study was partly supported by Grant-in-aid for scientific research (B) 21360462 from MEXT.

REFERENCES

1. M.Kondo, M.Takahashi, N.Sawada and K.Hata, *Corrosion of Steels in Lead-Bismuth Flow*, J.Nucl.Sci.Tech., **43**, 107(2006)
2. C.Fazio, G.Benamati, C.Martini, G.Palombarini, *Compatibility tests on steels in molten lead and lead-bismuth*, J.Nucl. Mater., **296**, 243 (2001)
3. Kenji Kikuchi, Kinya Kamata, Mikinori Ono, Teruaki Kitano, Kenichi Hayashi and Hiroyuki Oigawa, *Corrosion rate of parent and weld materials of F82H and JPCA steels under LBE flow with active oxygen control at 450 and 500 °C*, J.Nucl. Mater., **377**, 232 (2008)
4. Gupta, V., Argon, A.S., Cornie, J.A., Parks, D.M., *Measurement of interface strength by laser spallation technique*, J. Mech. Phys. Solids 40, 141 (1992)
5. Satou, M., Sato, T., Hasegawa, A. , *Role of Oxide Layer on Wall Thinning Caused by Liquid Droplet Impingement*, Proc. of PVP2009, ASME Pressure Vessels and Piping Division Conference (July 26-30, 2009 Prague, Czech Republic) PVP2009-77685.
6. Akamatsu, H., Satou, M., Sato, T., Jain, A., Gupta, V., Hasegawa, A., *Evaluation of bonding strength between yttria and vanadium alloys for development of self-cooled liquid blanket*, Proc. of ICFRM14, Int. Conf. Fusion Reactor Materials (Sept. 7-12, 2009, Sapporo, Japan)

Mater. Res. Soc. Symp. Proc. Vol. 1215 © 2010 Materials Research Society 1215-V16-24

Grain Boundary Characteristics Evaluation by Atomistic Investigation Methods

Yoshiyuki Kaji[1], Tomohito Tsuru[1] and Yoji Shibutani[2]
[1]Nuclear Science and Engineering Directorate, Japan Atomic Energy Agency, 2-4 Shirakata-Shirane, Tokai-mura, Ibaraki, Japan
[2]Department of Mechanical Engineering, Osaka University, 2-1 Yamadaoka, Suita, Osaka, Japan

ABSTRACT

The grain boundary has been recognized for one of the major defect structures in determining the material strength. It is increasingly important to understand the individual characteristics of various types of grain boundaries due to the recent advances in material miniaturization technique.

In the present study three types of grain boundaries: coincidence site lattice (CSL), small angle (SA), and random types are considered as the representative examples of grain boundaries. The grain boundary energies and atomic configurations of CSL are first evaluated by first-principle density functional theory (DFT) and the embedded atom method (EAM) calculations. SA and random grain boundaries are subsequently constructed by the same EAM and the fundamental characteristics are investigated by the discrete dislocation mechanics models and the Voronoi polyhedral computational geometric method. As a result, it is found that the local structures are well accorded with the previously reported high resolution-transmission electron microscope (HR-TEM) observations, and that stress distributions of CSL and SA grain boundaries are localized around the grain boundary core. The random grain boundary shows extremely heterogeneous core structures including many pentagon-shaped Voronoi polyhedral resulting from the amorphous-like structure.

INTRODUCTION

Interaction process of grain boundary and dislocation is one of the most important factors in the crystalline materials which contributes significantly to plastic behavior as well as the dislocation-dislocation process. Though these two lattice defects have been efficiently introduced in material strengthening well-known as Hall-Petch relationship [1, 2], they have the negative effects in the fracture behavior due to the stress concentration and segregation site at the boundary. Dislocations are successively piled up beside the grain boundary planes and they produce stress concentration which becomes a dominant stress factor of stress corrosion cracking (SCC). Ikuhara et al. observed detailed structure of low angle grain boundary in alumina bicrystals using HR-TEM, and analyzed the grain-boundary energies theoretically [3]. More recently, piled-up dislocations at the boundary and interaction between grain boundary and these dislocations under indentation-induced applied stress have been observed by in-situ nanoindentation test with TEM [4]. However, detailed mechanisms of the interaction between these processes and their influence on plastic behavior have not yet been definitively explored.

Computer simulations based on the atomistic models have been advanced for the last several decades and they allow us to treat the typical CSL and small-angle grain boundary structures. Basic properties in CSL grain boundaries restricted to the low sigma value such as $\Sigma 3$ and $\Sigma 5$ are implemented by DFT [5], and the effect of impurity atoms on these low sigma CSL are investigated [6, 7]. At the present time, however, DFT calculations have difficulty in handling grain boundary-dislocation system. Atomistic simulations by empirical potentials avoid this issue, and a large number of studies of grain boundary including dislocations are strenuously implemented. Rittner and Seidman investigated the equilibrium structures and corresponding grain boundary energies of various kinds of tilt CSLs. Swygenhoven et al. performed molecular dynamics simulations for grain boundary process of triple junction and dislocation emission under deformation [8, 9]. Yamakov et al. handled large scale parallel molecular dynamics simulations of polycrystal plasticity including dislocation-dislocation and dislocation-twin boundary reactions in nanocrystalline fcc metals [10]. On the other hand, polycrystalline materials used in the actual equipments include various types of grain boundaries, and therefore it is absolutely important to understand the relationship between basic features of grain boundaries and deformation behavior.

In the present study, we first focus attention on a huge variety of symmetric CSL tilt grain boundaries in fcc aluminum and copper. Grain boundary energies and excess free volumes of the equilibrium structures of these grain boundaries are investigated using empirical potentials. The fundamental characteristics of random grain boundary are investigated by the Voronoi polyhedral computational geometric method. Furthermore, the interaction between dislocation and small angle grain boundary is investigated by the discrete dislocation mechanics models.

FUNDAMENTAL GRAIN BOUNDARY PROPERTIES

We performed a sequence of calculations of structure stabilizations in fcc aluminum and copper to obtain the grain boundary energies of <110> tilt CSL grain boundary. Various kinds of CSL structures of $\Sigma 3$ to $\Sigma 99$ in both materials are constructed by supercell model with triaxial periodic boundary conditions, where the initial structures are constructed in accordance with the symmetrical rotation. Stable configurations are obtained by the conjugate gradient method as the total energy minimization scheme. Atomic configurations adjacent to the grain boundary planes are complicated, and therefore simple potential form have difficulty expressing these properties. The embedded atom method (EAM) type interatomic potential [11] is employed to express grain boundary structures, respectively. These potentials are proposed by Mishin et al. and the nonlinear properties related to the defect structure as well as elastic properties are greatly taken into account. Grain boundary energies and corresponding excess free volumes of CSL grain boundaries in relation to the misorientation angles are shown in Figure 1, where stable structures are calculated in the case that more than 70 atom layers in the direction normal to the grain boundary plane are given as the analysis model. Stress distribution driven from grain boundary is localized within several atom layers, and therefore the effect of boundary is negligibly small. In Figure 1 (a), $\Sigma 3$A and $\Sigma 11$A grain boundaries in both materials show extremely-low grain boundary energy known as energy cusps, where $\Sigma 3$A corresponds to the twin boundary. More specifically in aluminum, the energy cusps are presented in $\Sigma 3\{-111\}<110>$ and $\Sigma 11\{-113\}<110>$, and then grain boundary energies are calculated to be 75.0 mJ/m^2 and 150.5 mJ/m^2, respectively. Excess free volume in Figure 1 (b) can be evaluated by comparing the relative increase of volume to the unit area of grain boundary plane with the bulk volume in the single crystal: $(V_{GB}-V_{Bulk})/A_{GB}$, where V_{GB} is the volume including two grain boundary planes within the supercell, V_{Bulk} is the volume of the perfect single crystal and A_{GB} is the total area of grain boundary.   We found that they have two extremely small values at the angles related to Σ3 and Σ11. In addition to the energy cusps CSL grain boundary, excess free volume of other CSL grain boundaries tends to follow the corresponding grain boundary energy, i.e. there are correlative relationships between grain boundary energy and excess free volume. We therefore described these relationships in Figure 1(b) to determine the strength of the correlation, and significant correlations can be certainly confirmed in both materials. The first principle calculation results based on the DFT are also shown in Figure 1(a). It is found that EAM results are good agreement with DFT. The grain boundary energy of Σ3A, i.e. twin boundary has the similar relation to the stacking fault energy and the energies in the all grain boundary structures except the twin boundary of aluminum are smaller than those of copper.

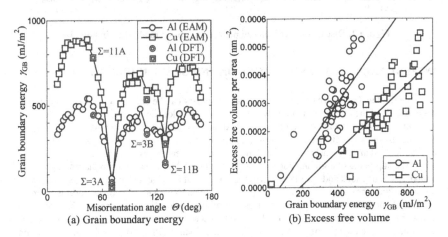

(a) Grain boundary energy   (b) Excess free volume

Figure 1. Grain boundary energies and excess free volumes of Σ3 to Σ 99 CSL grain boundary in Al and Cu.

On the other hand, it was clear that over 80% of irradiation assisted SCC (IASCC) occurred at random grain boundaries in the experimental SCC study [12]. Random grain boundary has lower symmetry contrary to the CSL grain boundary [13]. Therefore, understanding the physical properties of random grain boundary should be short cut to clarification of the mechanisms of SCC in the aged nuclear power plants. Even though the detailed structure of random grain boundary is not clear, at first, the voluntary angular grain boundaries are subsequently constructed by the EAM, and then we make the simulated random grain boundaries for aluminum by adding the twist on the normal axis of their grain boundary surface. The fundamental characteristics are investigated by the Voronoi polyhedral computational geometric method. Figure 2 shows distribution of tetragon, pentagon and hexagon around the random grain boundary. Many tetragonal- and hexagonal-shaped Voronoi polyhedra are existed in the grain by the reflection of fcc structure. On the other hand, a large number of pentagonal-shaped Voronoi polyhedra are especially existed around the grain boundary. The random grain boundary shows extremely heterogeneous core structures including many pentagonal-shaped Voronoi polyhedron resulting from the amorphous-like structure.

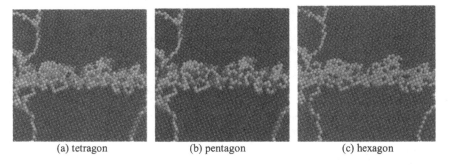

(a) tetragon   (b) pentagon   (c) hexagon

Figure 2. Distribution of Voronoi polyhedra around random grain boundary.

INTERACTION BETWEEN DISLOCATION AND GRAIN BOUNDARY

Since the local structure of small angle grain boundary can be expressed by the dislocations, we constructed the coupling model between the discrete dislocation method and the finite element method for the interaction between small angle grain boundary and dislocation. Figure 3(a) and (b) show the dislocation configuration of interaction between small angle grain boundaries and an edge dislocation for aluminum under shear stress conditions in the case I in which the slip plane of a gliding edge dislocation is located between two slip planes of grain boundary dislocations and case II in which the position of an edge dislocation exists on the same plane of a grain boundary dislocation, respectively. Even though an edge dislocation is trapped in the grain boundary dislocation in both cases, the gliding dislocation is trapped on the grain boundary surface in the case of Figure 3(a) and balanced near the grain boundary dislocation on the same plane in the case of Figure 3(b). The Peach-Koehler (PK) force for the escape of grain boundary dislocation in the different small angle grain boundaries are shown in Figure 3(c) and (d). The PK force for the escape in the maximum small grain boundary is not only depend on the position of dislocation, but also the angle, although the PK force is more localized in the case that the grain boundary dislocation is located on the different slip plane of the gliding dislocation. Concerning the influence of free surface, the stress distribution are shown in Figure 3(e) and (f) where the small angle grain boundary is existed in the bulk and near the free surface, respectively. For the small angle grain boundary in the bulk, it is found that the stress distribution is concentrated locally and more localized with increase of angle. On the other hand, in the case that the small grain boundary is existed near the free surface, it is confirmed that the traction-free condition achieved by the superposition near the free surface and the stress distribution represents far-reaching consequence because the individual stress field is not contradicted each other. Furthermore, it is seen that the trapping force of the grain boundary dislocation becomes small with approaching the free surface from the similar analysis presented in Fig. 3 (C) and (D).

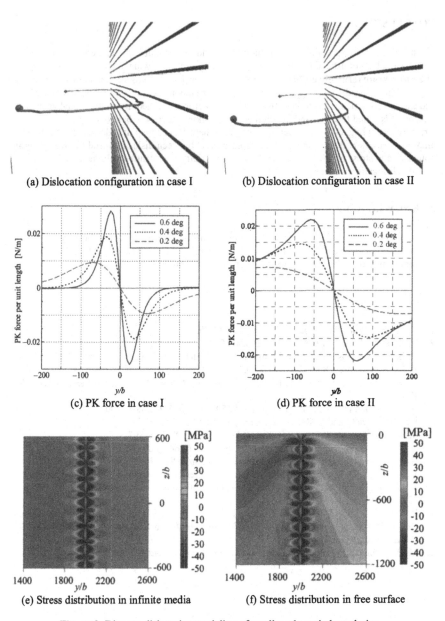

(a) Dislocation configuration in case I

(b) Dislocation configuration in case II

(c) PK force in case I

(d) PK force in case II

(e) Stress distribution in infinite media

(f) Stress distribution in free surface

Figure 3. Discrete dislocation modeling of small angle grain boundaries.

CONCLUSIONS

Fundamental properties of the CSL tilt grain boundaries in aluminum and copper are investigated by atomistic simulations. Grain boundary energy of wide range of CSL structures of $\Sigma 3$ to $\Sigma 99$ are relatively evaluated, and two energy cusps are reproduced in both aluminum and copper. Moreover, there are approximately linear correlations between grain boundary energy and excess free volume. From investigation by the Voronoi polyhedral computational geometric method, it is found that the random grain boundary shows extremely heterogeneous core structures including a large number of pentagonal-shaped Voronoi polyhedra resulting from the amorphous-like structure. Furthermore, the interaction between dislocation and small angle grain boundary and stress field generated by the grain boundary are quantitatively modelled by the discrete dislocation mechanics.

ACKNOWLEDGMENTS

The authors T. T. and Y. S. acknowledge financial support from the Japan Society for the Promotion of Science (JSPS), Grants-in-Aid for Scientific Research (S, Grant No. 20226004) and Grant-in-Aid for Young Scientists (B, Grant No. 21760090).

REFERENCES

1. E. O. Hall, *Proc. Phys. Soc. B*, 64, (1951), 747-753.
2. N. J. Petch, *J. Iron and Steel Inst.*, (1953), 25–28.
3. Y. Ikuhara, H. Nishimura, A. Nakamura, K. Matsunaga, T. Yamamoto and K. P. D. Lagerlof, *J. Am. Ceram. Soc.*, 86, (2003), 595-602.
4. A. M. Minor, E.T. Lilleodden, E. A. Stach and J. W. Morris, Jr., *J. Mater. Res.*, 19-1, (2004), 176-182.
5. A. F. Wright and S. R. Atlas, *Phys. Rev. B*, 50-20 (1994), 15248-15260.
6. J. S. Braithwaite and P. Rez, *Acta Mater.*, 53, (2005), 2715–2726.
7. M. Yamaguchi, M. Shiga and H. Kaburaki, *Science*, 307, 393-397.
8. H. Van Swygenhoven and P. M. Derlet, *Phys. Rev. B*, 64, (2001), 224105.
9. H. Van Swygenhoven, P. M. Derlet and A. Hasnaoui, *Phys. Rev. B*, 66 (2002), 024101.
10. V. Yamakov, D. Wolf, S. R. Phillopot and H. Gleiter, *Acta Mater.*, 51, (2003), 4135-4147.
11. Y. Mishin, D. Farkas, M. J. Mehl and D. A. Papaconstantopoulos, *Phys. Rev. B*, 59, (1999), 3393-3407.
12. Y. Miwa, Y. Kaji, T. Tsukada, Y. Kato, T. Tomita, N. Nagata, K. Douzaki and H. Taguchi, *Proc. 13th International Conference on Environmental Degradation of Materials in Nuclear Power Systems* (2007)
13. M. P. Allen and D. Tildesley, *Computer Simulation of Liquid*, Clarendon, Oxford (1987)

Mater. Res. Soc. Symp. Proc. Vol. 1215 © 2010 Materials Research Society 1215-V16-26

Characterization of Changes in Properties and Microstructure of Glassy Polymeric Carbon Following Au Ion Irradiation

Malek Abunaemeh[1,2*], Mohamed Seif[3], Young Yang[4], Lumin Wang[5], Yanbin Chen[5], Ibidapo Ojo[1,2], Claudiu Muntele[1] and Daryush Ila[1,2]

[1]Center for Irradiation of Materials, Alabama A&M University Research Institute, Normal, AL 35762
[2]Physics Department, Alabama A&M University, Normal, AL 35762
[3]Mechanical Engineering Department, Alabama A&M University, Normal, AL 35762
[4]Engineering Physics Department, University of Wisconsin, Madison, WI, 53706
[5]Department of Nuclear Engineering & Radiological Sciences, University of Michigan, Ann Arbor, MI 48109-2104

* Corresponding author

Abstract

The TRISO fuel has been used in some of the Generation IV nuclear reactor designs [1,2]. It consists of a fuel kernel of UO_x coated with several layers of materials with different functions. Pyrolytic carbon (PyC) is one of the materials in the layers. In this study we investigate the possibility of using Glassy Polymeric Carbon (GPC) as an alternative to PyC. In this work, we are comparing the changes in physical and microstructure properties of GPC after exposure to irradiation fluence of 5 MeV Au equivalent to a 1 displacement per atom (dpa) at samples prepared at 1000, 1500 and 2000°C. The GPC material is manufactured and tested at the Center for Irradiation Materials (CIM) at Alabama A&M University. Transmission electron microscopy (TEM), Rutherford backscattering spectroscopy (RBS), X-ray photoelectron spectroscopy (XPS) and Raman spectroscopy were used for the analysis.

INTRODUCTION

Current high temperature gas cooled rectors are designed with the use of coated fuel particles that are dispersed in a graphite matrix to form fuel elements called Tristructural-isotropic (TRISO) [1,2]. As seen in figure 1[3], TRISO consists of a microspherical kernel of uranium oxide (UO_x) coated with a layer of porous carbon buffer surrounding it to contain any particle dimensions changes or gas buildup. This layer is followed with an inner pyrolitic carbon (PyC) followed by a layer of silicon carbide (SiC) followed by an outer PyC layer.

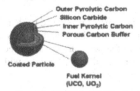

Outer Pyrolytic Carbon
Silicon Carbide
Inner Pyrolytic Carbon
Porous Carbon Buffer

Coated Particle

Fuel Kernel
(UCO, UO₂)

Figure 1: TRISO fuel layers structure

The pebbles or the graphite blocks containing the TRISO fuel is directly immersed in the cooling fluid that extracts the heat outside of the reactor core while keeping the inside within the operational temperature limits [4,5]. If the fissile fuel is in direct contact with the cooling fluid, there are great chances that radioactive fission fragments will be carried out of the reactor core and contaminate all other equipment [6]. Therefore, in order to minimize such leaks, TRISO is designed with PyC as the diffusion barrier material.

GPC is widely used for various applications from artificial heart valves to heat-exchangers and other high-tech products that are developed for the space and medical industries. This lightweight material can maintain dimensional and chemical stability in adverse environment and very high temperatures (up to 3000°C).

The primary purpose of this study is to understand the changes in fundamental properties (chemical and mechanical stability) of GPC prepared at different temperatures (1000, 1500 and 2000°C). Each one of these samples was bombarded with 5MeV Au to an effect of 1 DPA. More generally, we are trying to understand the fundamental mechanisms of defect creation in sp^2-bonded carbons under particle radiation. The type and concentration of such defects have deep implications in the physical properties of carbon-based materials for various applications from carbon-based quantum electronics to structural components for aerospace applications. Here, in particular, the specific application is in an extreme radiation environment, the core of nuclear reactor. This study will help to determine GPC eligibility for future irradiation testing for a specific application in an extreme radiation environment in the nuclear reactor. It will also help to determine if GPC will be a good choice as a diffusion barrier in the TRISO fuel that will be used in the next generation of nuclear reactors

EXPERIMENTAL DETAILS AND DISCUSSION

It was very important to characterize the GPC samples before irradiation to ensure that there was no contamination in the sample. XPS and RBS (not shown in this paper) confirmed that. The first step was to develop a procedure for making the GPC sample. This was a lengthy process. It starts by taking phenolic resin that is placed in a beaker and in a sonication bath for about 1-2 hours. The resin is then taken out of the sonication bath and is poured carefully in the desired mold. The mixture is then heated slowly at a rate of 20°C/day for 6 days to lower the possibility of any bubble forming during the gelling stage. The hardened gel is removed from the mold and heated to temperatures up to 2000°C slowly over the course of five days. At this stage they are fully pyrolized GPC samples. The heating profile that was developed for GPC in our lab was followed during this process [7,8]. This developed heating profile can be used to heat samples up to 2500°C. The sample is then left to cool off in the furnace until it reaches room temperature. Figure 2 shows a plot of the heating profile that was followed to prepare the samples.

Several methods and instruments were used to characterize the elemental and mechanical properties for our GPC samples including XPS, RBS, TEM and Raman spectroscopy. The three samples that were prepared at three different temperatures (1000, 1500, and 2000°C) were then irradiated with 5MeV Au with a fluence of 1.3×10^{16} ions/cm^2, to an effect of 1 dpa

146

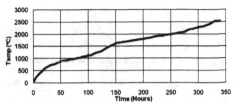

Figure 2: Heating profile used for the GPC samples in graphite furnace

The first technique used to analyze the GPC samples was XPS. XPS is a surface chemical analysis technique used to identify the elemental and chemical components that make up the sample. The sample is irradiated with a beam of X-rays (Al K_α). The number of electrons at each kinetic energy (KE) value is then counted. Figure 3 shows the XPS spectrum for the carbon lines that are associated with in the GPC samples prepared at different temperatures (1000, 1500 and 2000°C). Figure 4 shows the oxygen lines that appeared within the three GPC samples. The Oxygen lines in the XPS spectra are due to the hydrocarbon contamination of the sample surface from the oil vapors in the vacuum system.

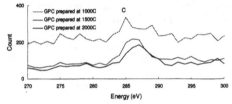

Figure 3: XPS Spectrum for the C lines in the GPC samples prepared at (1000, 1500 and 2000°C)

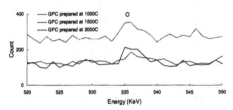

Figure 4: XPS Spectrum for the O lines in the GPC samples prepared at (1000, 1500 and 2000°C)

TEM was also used for characterizing the microstructures of GPC samples prepared at three different temperatures as seen in Figure 5. TEM was done at the Department of Nuclear Engineering & Radiological Sciences at the University of Michigan.

147

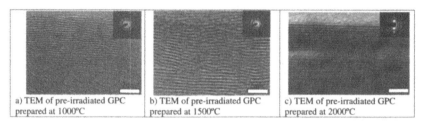

| a) TEM of pre-irradiated GPC prepared at 1000°C | b) TEM of pre-irradiated GPC prepared at 1500°C | c) TEM of pre-irradiated GPC prepared at 2000°C |

Figure 5: TEM comparison of pre-irradiated GPC samples prepared at various temperatures

Figure 5(a-c) are typical HRTEM images of GPC prepared at 1000, 1500 and 2000°C. All of three samples show the graphite-like layered structure. There is progressively more alignment of layers with temperature. Based on these results, higher processing temperatures would be more desirable for obtaining more isotropic GPC material. The crystalline quality of GPC processed at 2000°C is closer to single crystalline graphite compared with other two samples. This conclusion was supported by the electron energy loss spectroscopy (EELS) result shown in figure 6. The π peak is proportional to sp2 bonding (graphite-like) percentage in C specimens. The π peak at GPC processed at 2000°C is higher than the other two samples. GPC processed at 2000 °C is more graphitized.

RBS (not shown) was also done to confirm that there was no contamination on the sample prior to Au irradiation. A pure classical C^{12} was seen on the GPC samples.

Figure 6: EELS of pre-irradiated GPC samples at various temperatures.

The GPC samples irradiation took place at the Center for Irradiation of Materials (CIM). The samples were then irradiated with 5MeV Au with a fluence of 1.3×10^{16} ions/cm^2 to an effect of 1 DPA.

TEM on the post irradiated samples were done at the Engineering Physics Department at the University of Wisconsin. Figure 7 shows that the gold distribution was about 1.5 μm inside the GPC, matching the SRIM simulation as seen in figure 8. Raman Spectroscopy was used to monitor the changes in the chemical bonding for the GPC samples. Figure 9 shows the Raman spectroscopy for the pre irradiated GPC samples while Figure 10 shows the irradiated samples. The D (distorted) lines and the G (graphitic) lines show the sample is destroyed after irradiation. Table 1 shows the D and G calculation before and after irradiation. The intensity of the D band in Raman depends on the size of the graphite microcrystals in the sample.

Figure 7: TEM of Au irradiated GPC

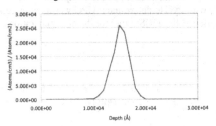

Figure 8: SRIM Simulation of Au irradiated GPC

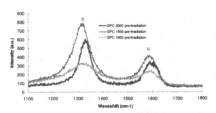

Figure 9: Raman spectroscopy of pre-irradiated GPC samples

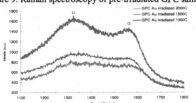

Figure 10: Raman spectroscopy of the post-irradiated GPC samples

	GPC prepared at 1000°C	GPC prepared at 1500°C	GPC prepared at 2000°C
D/G of pre Au irradiation	1.87	1.65	1.75
D/G of post Au irradiation	4.8	3.51	2.25

Table 1: D/G ratio for pre and post irradiated GPC samples

CONCLUSIONS

TEM of GPC samples fabricated at 1000, 1500 and 2000°C showed that higher fabrication temperatures produce more ordered, graphite-like, layered structure samples and higher processing temperatures may be better for obtaining more isotropic GPC material.TEM also showed that Au penetrated 1.5 μm deep into the sample, which matched the SRIM simulation. Raman showed that the samples were completely damaged. The next step is to implant Ag and other fission product elements to GPC samples prepared at (1000, 1500 and 2000°C) and to compare the results of the diffusivity of these elements in GPC with PyC to determine if GPC was a better barrier for fission products than PyC to be used in the TRISO fuel for the next generation of nuclear reactors.

ACKNOWLEDGMENTS

This research was supported and funded by the AAMRI Center for Irradiation of Material, NSF Alabama GRSP EPSCOR, and DoE NERI-C project number DE-FG07-07ID14894.

REFERENCES

1. Association of German Engineers (VDI), the Society for Energy Technologies (publ.) (1990). *AVR - Experimental High-Temperature Reactor, 21 Years of Successful Operation for A Future Energy Technology*. Association of German Engineers (VDI), The Society for Energy Technologies. pp. 9–23. ISBN 3-18-401015-5
2. K. Fukuda, T. Ogawa, K. Hayashi, S. Shiozawa, H. Tsuruta, I. Tanaka, N. Suzuki, S. Yoshimuta, M. Kaneko, J. Nucl. Sci. Technol. 28 (1991) 570.
3. www.nextgenerationnuclearplant.com/i/triso.gif
4. D. Olander *J. Nucl. Mater.* 389 (2009) 1-22
5. http://www.Iaea.org, HTGR Knowledge Base
6. D. A. Petti, J. Buongiorno, J. T. Maki, R. R. Hobbins and G. K. Miller, "Key differences in the fabrication, irradiation and high temperature accident testing of US and German TRISO-coated particle fuel, and their implications on fuel performance". Nuclear Engineering and Design, Volume 222, Issues 2-3, June 2003, Pages 281-297
7. Jonathan Fisher, Dec 1996, "Active Crucible Bridgman system: Crucible characterization, RF Inductor Design And Model", Dissertation, Alabama A&M University
8. H Maki, L Holland, G Jenkins and R Zimmerman, "Maximum heating rates for producing undistorted glassy carbon we determined by wedge-shaped samples", J. Mater. Res., Vol 11, No 9, Sep 1996, pages 2368-2375

Mater. Res. Soc. Symp. Proc. Vol. 1215 © 2010 Materials Research Society   1215-V16-34

Thermal Conductivities of Cs-M-O (M = Mo or U) Ternary Compounds

Kazuyuki TOKUSHIMA[1]   Kosuke TANAKA[2]   Ken KUROSAKI[1]
Hiromichi GIMA[1]   Hiroaki MUTA[1]   Masayoshi UNO[3]
Shinsuke YAMANAKA[1,3]
1 Division of Sustainable Energy and Environmental Engineering, Graduate School of
Engineering, Osaka University, 2-1 Yamadaoka, Suita 565-0871, Japan
2 Japan Atomic Energy Agency, Narita-cho 4002, Oarai-machi, Higashiibaraki-gun, Ibaraki,
311-1393, Japan
3 Research Institute of Nuclear Engineering, Fukui University, 3-9-1 Bunkyo, Fukui 910-8507,
Japan

ABSTRACT

Thermal conductivities of Cs-M-O (M = Mo or U) ternary compounds, observed in the pellet-cladding gap region and in the pellet periphery in irradiated oxide fuels with high oxygen potentials, were investigated. Bulk samples of Cs_2MoO_4 and Cs_2UO_4 were prepared by hot pressing or spark plasma sintering, and their thermal diffusivities were measured by the laser flash method from room temperature to 823 K for Cs_2MoO_4 and to 900 K for Cs_2UO_4. The thermal conductivities were evaluated from the thermal diffusivity and bulk density, and the specific heat capacity values available in the literature. The thermal conductivities of Cs_2MoO_4 and Cs_2UO_4 were quite low compared with UO_2 (e.g. 0.5 $Wm^{-1}K^{-1}$ at 800 K for Cs_2MoO_4).

INTRODUCTION

It is important to understand the behavior of fission products (FPs) to evaluate fuel performance. For example, in high burn-up oxide fuels, some FPs dissolve in the fuel matrix and others form oxide or metallic inclusions, both of which would affect the physical and chemical properties of the fuels. Among FPs, cesium is one of the most important due to its high fission yield in mixed oxide (MOX) fuel when burned in the fast reactor. In the post irradiation examinations of the Phenix fuel pins, accumulation of cesium molybdate in the fuel-cladding gap was observed [1]. Cesium uranates were generally observed in the fuel pellet periphery. In order to improve the accuracy of prediction of the thermal performance of the MOX fuel to high burn-up, it is indispensable to measure the thermal conductivity of these cesium compounds.

Many compounds have been identified in the Cs-Mo-O and Cs-U-O system and their crystal structures and thermochemical data have been reviewed and summarized [2]. Experimental thermal conductivity data of their compounds is restricted however to Cs_2MoO_4 [3,4] and Cs_2UO_4 [5] .

The objective of the present study was to acquire data on thermal conductivities of Cs_2MoO_4 and Cs_2UO_4 using samples fabricated by hot press and spark plasma sintering (SPS) technique.

EXPERIMENTAL

Powders of Cs_2MoO_4 (Mitsuwa Chemicals; purity 99 %), Cs_2CO_3 (Kishida Chemicals; purity 99.99 %) and U_3O_8 were prepared as raw materials. Cs_2UO_4 powder was obtained by the following procedure: Cs_2CO_3 and U_3O_8 powders were weighed out as their hyperstoichiometrical amounts (U_3O_8/Cs_2CO_3 molar ratios were 1/6) and mixed thoroughly in an agate mortar with a pestle. The mixture was pressed into disks at 150 MPa pressure and room temperature for 5 min. The disks were placed in an alumina boat and heated at 873 K for 10h in flowing dry air.

Cs_2MoO_4 powder was packed into a graphite die and pressed and sintered in the SPS apparatus (SPS-515S, Sumitomo Coal Mining, Japan) under the following conditions: 30 MPa; Ar flow; 873 K; 10 min. The increase rate of temperature was about 50 K/min. The temperature on the surface of the graphite die was measured by an optical pyrometer. Cs_2UO_4 powder was packed into the graphite die and pressed and sintered by the SPS and hot press (N2011-00, Motoyama Co., Ltd., Japan) techniques. For the SPS technique, the powder was pressed and sintered under the conditions: 30 MPa; Ar flow; 873 K; 20 min. The increase rate of temperature was about 50 K/min. For the hot press technique, the powder was pressed and sintered under the conditions: 50 MPa; Ar flow; 873 K; 3 h. The increase rate of temperature was about 10 K/min. The temperature on the surface of the graphite die was measured by a thermocouple.

The crystal structure and phases were determined by powder X-ray diffraction (XRD) analysis using Cu-K_{α} radiation. Because of the hygroscopicity of Cs_2MoO_4, the powder for the XRD was mixed with epoxy resin in the sample holder. The lattice parameter was calculated by Cohen's method from the measured XRD pattern. The differential thermal analysis of Cs_2MoO_4 powder was measured by the thermogravimetry-differential thermal analysis (TG-DTA) apparatus (Thermo plus TG8120, Rigaku Co., Ltd., Japan).

The thermal diffusivities of the samples were measured by a laser flash method in the range from room temperature to 823 K for Cs_2MoO_4 and to 900 K for Cs_2UO_4 in vacuum using the TC-7000 apparatus (ULVAC Co.). Since the colors of the disk-shaped samples were white for Cs_2MoO_4 and orange for Cs_2UO_4, their surfaces were coated with gold and carbon. The thickness of the disk samples was about 1 mm. The measurement was performed three times at each temperature and the average value was taken as the measured value. The thermal conductivity (κ) was determined by the equation:

$$\kappa = \alpha C_p \rho \qquad (1)$$

where α is thermal diffusivity, C_p is the specific heat capacity, and ρ is the experimental density at room temperature of the sample. The values of the specific heat capacity were obtained from the literature [6,7]. The characteristics of the disk samples are given in Table 1.

152

RESULTS AND DISCUSSION

Figure 1 shows the XRD patterns of Cs_2MoO_4 powder samples fixed with epoxy resin before and after SPS. The pattern of the Cs_2MoO_4 powder before SPS was not identical with the data in the JCPDS file [8] but the pattern after SPS was. The difference for the former was caused by the moisture uptake the powder before it was sintered. The broad peak seen at low angle can be attributed to the epoxy resin. The XRD patterns of the Cs_2UO_4 powder before sintering are shown in Figure 2 with JCPDS file patterns [9]. Impurity peaks were found for the powder U_3O_8/Cs_2CO_3 mixed with the molar ratio= 1/3, but the XRD pattern of the powder mixed hyperstoichiometrically, U_3O_8/Cs_2CO_3 molar ratio = 1/6, was identical with the data in the JCPDS file. These impurity peaks were attributed to α-$Cs_2U_2O_7$ and β-$Cs_2U_2O_7$. Their presence was also mainly due to the hygroscopicity of Cs_2CO_3.

Table 1. Characteristics of disk samples

Sample	Sintering method	Lattice parameter (nm)			Theoretical density (g/cm^3)	Relative density (%T. D.)
		a	b	c		
Cs_2MoO_4	SPS	6.563	11.618	8.514	4.36	94.9
Cs_2UO_4-1	Hot pressing	4.400		14.831	6.57	77.2
Cs_2UO_4-2	SPS	4.392		14.833	6.59	79.3

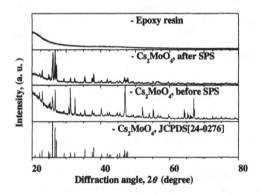

Figure 1. XRD patterns of Cs_2MoO_4 before and after SPS together with literature data [8].

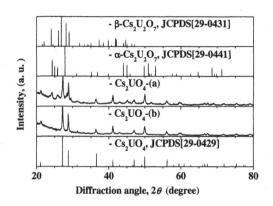

Figure 2. XRD patterns of Cs_2UO_4 (U_3O_8/Cs_2CO_3 molar ratios were 1/3 (a), 1/6 (b)) together with literature data [9].

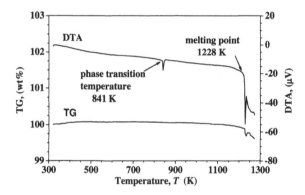

Figure 3. Differential thermal analysis of Cs_2MoO_4 powder.

Figure 3 shows measured results of TG-DTA. The phase transition was found at 841 K and the melting point was 1228 K, which agreed with the reported values [6]. Cs_2MoO_4 has an orthorhombic structure at room temperature and has a phase transition to a hexagonal structure at 841 K. The thermal conductivities of the disk samples evaluated by Eq. (1) are shown in Figure 4 as a function of temperature together with reported data [4,5]. The three data sets of the thermal conductivities in this study denote the same tendency as past research.

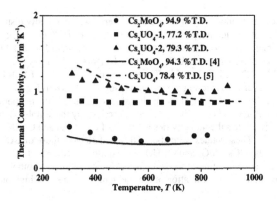

Figure 4. Thermal conductivities of Cs_2MoO_4 and Cs_2UO_4, together with reported values [4,5] as a function of temperature.

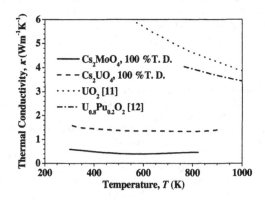

Figure 5. Comparison of corrected thermal conductivities of Cs_2MoO_4, Cs_2UO_4, UO_2 [11] and $U_{0.8}Pu_{0.2}O_2$ [12] as a function of temperature.

To obtain the thermal conductivity of Cs_2MoO_4 and Cs_2UO_4 with 100 % theoretical density, the present data were corrected by the analytical equation of Schulz [10]:

$$\kappa_p = \kappa_0 \, (\, 1 - P \,)^{1.5} \tag{2}$$

155

where κ_p is the thermal conductivity of the porous material with randomly distributed spherical pores, κ_0 is the thermal conductivity of the material with the 100 %T.D., and P is the porosity. The corrected thermal conductivities for Cs_2MoO_4 and Cs_2UO_4 with 100 % T.D. are shown in Figure 5. Thermal conductivity of Cs_2UO_4 with 100 % theoretical density was the average of Cs_2UO_4 with 77.2 % T.D. and 79.3 % T.D. The thermal conductivities of UO_2 [11] and MOX fuel ($U_{0.8}Pu_{0.2}O_2$) [12] are also presented for comparison. As shown in Figure 5, the thermal conductivities of Cs_2MoO_4 and Cs_2UO_4 were almost independent of the temperature, indicating that the mechanism of thermal conduction was not described by the simple phonon conduction theory. The respective thermal conductivities at 800 K of Cs_2MoO_4 and Cs_2UO_4 were 0.5 $Wm^{-1}K^{-1}$ and 1.4 $Wm^{-1}K^{-1}$. These values were about 10 % and 30 % of those of UO_2. Because the thermal conductivities of Cs_2MoO_4 and Cs_2UO_4 were quite low compared with UO_2 and MOX fuel, the effect of the accumulation of Cs-M-O (M = Mo or U) ternary compounds should be taken into consideration for the evaluation of the fuel temperature in the high burn-up region.

CONCLUSIONS

The thermal diffusivities of Cs_2MoO_4 and Cs_2UO_4 using samples fabricated by hot press and SPS techniques were measured by a laser flash method in the range from room temperature to 823 K for Cs_2MoO_4 and to 900 K for Cs_2UO_4. The thermal conductivities of these cesium ternary oxides were quite low compared with UO_2 and MOX fuel. This is consistent with previous findings. These results would be useful for evaluating the thermal performance of MOX fuels at the high burn-up region in the fast reactors.

REFERENCES

[1] M. Tourasse, M. Boidron and B. Pasquet, *J. Nucl. Mater.*, 188 (1992) 49.
[2] E. H. P. Cordfunke, R. J. M. Konings, *Thermochemical Date for Reactor Materals and Fission Products*, Elsevier, Amsterdam, 158 (1990).
[3] K. Minato, M. Takano, K. Fukuda, S. Sato, H. Ohashi, *J. Alloys Comp.*, 255 (1997) 18.
[4] T. Ishii, T. Mizuno, *J. Nucl. Mater.*, 231 (1996) 242.
[5] M. Takano, K. Minato, K. Fukuda, S. Sato, H. Ohashi, *J. Nucl. Sci. and Tech.*, 35 (1998) 485.
[6] R. J. M. Konings, E. H. P. Cordfunke, *Thermochim. Acta*, 124 (1988) 157.
[7] D. R. Fredrickson, P. A. G. O'Hare, *J. Chem. Thermodynamics*, 8 (1976) 353.
[8] JCPDS [24-0276], F. X. N. M. Kools, A. S. Koster, G. D. Rieck, *Acta Cryst.*, 26 (1970) 1974.
[9] JCPDS [29-0429], [29-0431], [29 0441], A. B. Van Egmond, Thesis, University of Amsterdam, (1970).
[10] B. Schulz, *KfK-1988*, (1974).
[11] P. G. Lucuta, *J. Nucl. Mater.*, 223 (1995) 51.
[12] D. G. Martin, *J. Nucl. Mater.*, 110 (1982) 73.

Mater. Res. Soc. Symp. Proc. Vol. 1215 © 2010 Materials Research Society   1215-V16-44

In situ Characterization of UO₂ Microstructure Changes During an Annealing Test in an Environmental Scanning Electron Microscope

M. Marcet[1], Y. Pontillon[1], L. Desgranges[1], D. Simeone[2], I. Aubrun[1], I. Felines[1], L. Brunaud[1]

[1] Commissariat à l'Energie Atomique DEN/CAD/DEC, F-13108 Saint-Paul-lez-Durance, France
[2] Commissariat à l'Energie Atomique DEN/DANS/DMN, F-91191 Gif sur Yvette, France
CNRS Ecole Centrale Paris SPMS MFE, F-92290 Châtenay-Malabry, France

ABSTRACT

A 1 µg High Burn Up Structure (HBS) fragment was extracted from a UO₂ fuel pellet irradiated for 7 cycles in a EDF Pressurised Water Reactor (PWR). In situ examinations were performed with an Environmental Scanning Electron Microscope (ESEM) in order to characterize UO₂ microstructure evolution during a temperature ramp up to 1,600K. The results are compared to previously published data on HBS annealing tests performed in a Knudsen cell where observed burst releases are explained as sample cracking during the experimental sequence.

INTRODUCTION

After irradiation in normal operating conditions, used UO₂ fuel exhibits different microstructure as a function of the radial position in the fuel pellet. These microstructures reveal the thermal gradient existing in the fuel pellet during operation, which is due to homogeneous production of heat by fission in the pellet. More specifically the High Burn-up Structure (HBS) appears at the outer rim of the pellet, its coldest part, when the local burn-up reaches a 60-70 GWd/tM threshold [1-4]. This structure is characteristic with the subdivision of initial grains into small grains and the presence of important porosities containing pressurised gas bubble. The fission gases behaviour during a temperature transient is a key issue regarding the PWR's nuclear fuel licensing [5]. This behaviour is generally studied in hot cells by measuring the fission gas release out of a slice of used nuclear fuel as a function of a predetermined thermal history [6-7].

Recently the fission gas release of the HBS was studied in a Knudsen cell [8]. These releases are continuous processes except the main steps at 1,000K and 1,500K, which are characterized by explosive gas release evidenced from the pressure transients observed by the vacuum gauges. These pressure spikes are explained by burst releases of the gas contained in closed pores. Other annealing test performed in our team on a whole fuel pellet [9] confirmed that the HBS released gases at approximately the same temperature. Besides some mechanical degradation of the HBS was observed after the thermal sequence.

In this study we used Environmental Scanning Electron Microscope coupled with an heating stage in order to characterize the change of the UO₂ microstructure during temperature ramps in relation to the burst release. The gas release mechanism will be discussed as a function of the cracks observed both at the surface and in the bulk.

EXPERIMENTAL SECTION

1. Fuel samples

A slice taken out of a UO_2 fuel rod irradiated up to 77 GWd/t (7 cycles) in a PWR operated by EDF was used in order to collect a fragment of HBS. The main manufacturing and irradiation characteristics of this fuel rod are presented in the Table 1.

Table 1. Manufacturing and irradiation characteristics of the fuel rod

Fuel	Cladding	Enrichment	No. of irradiated cycles	Burn-Up [MWd/t]
UO_2	M5	4.5% ^{235}U	7	76 941

In this slice the HBS consists in a 150 μm wide layer located at the outer rim of the pellet with a local burn-up of around 140 GWd/t. Optical characterisation on adjacent sample evidenced that this HBS was stuck to the inner face of the zirconium alloy cladding. As a consequence HBS is still located on the inner face of the cladding after the fuel pellet is removed out of the fuel rod slice which allowed the collection of HBS fragments in hot cell with a specific method. The fuel slice is first cut parallel to the rod axis in order to remove the fuel pellet. The remaining cladding is set in a vice as shown in Figure 1. Then a metallic sample holder in molybdenum is tightened on the cladding in order to incrust the HBS fragment (see Figure 1). This sample holder had a sufficiently low radioactivity to be sent from the hot cells to the ESEM.

Metallic sample holder   Cladding with HBS

Figure 1. HBS fragment sampling in a vice

The HBS fragment we studied weights around 1μg and measures 30μm × 60 μm as observed by SEM on Figure 4a. This fragment corresponds to a HBS zone which is fully restructured with a small grain size (< 1 μm) and large pores (> 1 μm).

2. ESEM experiments

In situ examinations of UO_2 microstructure changes during annealing tests were carried out using an Environmental Scanning Electron Microscope (ESEM FEG XL30, Philips) equipped with an heating stage and performed under high purity nitrogen atmosphere. The HBS fragment was observed on its sample holder during temperature ramp which the cycle is detailed Figure 2. Images in environmental mode were taken all along this thermal cycle. The quality of these images depends on the residual atmosphere used in the environmental mode. Because of the thermal cycle, the residual atmosphere was modified during the experiment which required continuous tuning of the ESEM . High quality images were also taken before and after the annealing treatment in order to observe the changes in the fragment morphology due to temperature.

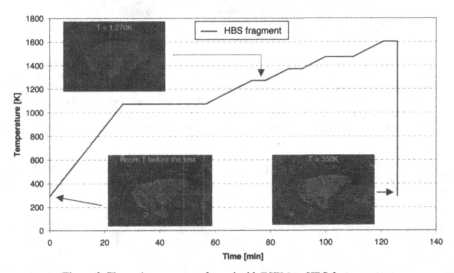

Figure 2. Thermal sequence performed with ESEM on HBS fragment

RESULTS AND DISCUSSION

Another slice was cut near to the one used for ESEM experiment. It underwent an annealing test in MERARG facility [9]. The Fission gas released out of the slice was measured during the annealing test. Fission gas release was observed at 900K and 1,450K. Furthermore, some mechanical degradation of the HBS was observed by SEM after the thermal sequence. These results are consistent with the one reported by Hiernaut et al. [8] concerning the fission product release and microstructure changes during laboratory annealing of a high burn-up fuel on a Knudsen cell under vacuum. Specific FGRs occur at 1,000K and 1,500K which are characterized by explosive gas release evident from the pressure transients observed on the vacuum gauges. These pressure spikes are explained by burst releases of the gas contained in closed pores. The release burst at 1500K causes the fracture of the sample like it is observed Figure 3. These experience was performed on much larger samples (≈ 800 μm width) than ours and at higher burn-up (average burn-up of 200 GWd/t). At the end of the experience, the fragment average width is about 50 μm.

Figure 3. SEM observation of sample fracture at 1,500K by Hiernaut et al. [8]

According to the interpretation given by Hiernaut, mechanical degradation of the HBS fragment due to burst releases of the gas contained in closed pores was expected to occur during the ESEM experiment. The images obtained on the HBS fragment at room temperature before and after the thermal sequence (Figure 4) did not evidence any sudden mechanical degradation of the HBS fragment all along the thermal sequence.
Figure 5 shows in situ images performed on HBS fragment at different steps of the temperature cycle.

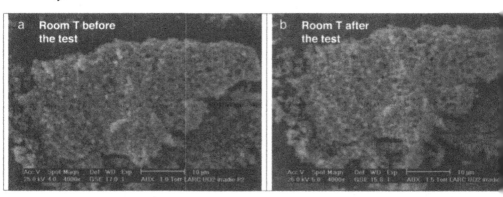

Figure 4. Comparison of ESEM images (× 4,000) of HBS fragment
a) before and b) after the annealing test at the room temperature

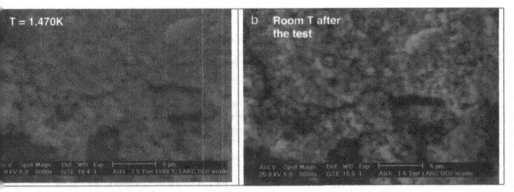

Figure 5. In situ examinations by ESEM on HBS fragment (× 8,000)
a) at 1,470K and b) at the room temperature after the test

Some small changes before and after the test are nevertheless observed. The comparison of images before, after and during the experiment (Figures 4 - 5) shows a micro structural evolution:

   i)   the grains are more highlighted at the room temperature than at 1,470K,
   ii)   it appears an opening of pre-existing cracks and micro cracks,
   iii) some porosities are revealed at room temperature (at right top)

In the same zone, the comparison between the images taken at room temperature before the test and the image at 1,470K shows an increase of the grain size and the porosities.
The thermal expansion and the thermal attack effect cause the rounding of the grains by revealing the microstructure and a closure of micro cracks during the heating up until 1,370K. The temperature range between 1,370K and 1,470K affects the microstructure by a "re"opening of micro cracks which participates to the FGR.

The difference of behaviour of the HBS in ESEM experiment compared to Hiernaut's in Knudsen cell can't be explained by the experimental conditions applied. The temperatures reached during ESEM experiment are over the temperatures at which Hiernaut observed burst release (around 1,500K). Moreover this temperature is very closed to the temperature at which gas release was observed by annealing test on the same fuel than the one used in ESEM experiment. The difference between residual atmosphere of ESEM and vacuum in Knudsen cell is not thought to influence the gas release of HBS because both did not lead to any significant oxidation of the HBS sample.

The major difference between our experiment and Hiernaut's one concerns the HBS sample itself, which has not the same burn-up and size. In Hiernaut's experiment samples are: i) 10 times bigger and ii) the burn-up is higher, it affects the microstructure with more porosities and higher gas bubble pressures. Thus fractures are likely to be initiated more easily in Hiernaut's

161

samples. The sample pulverization which occurs at 1,500K creates small fragments which average size corresponds to our fragment size.
Using the comparison between our results and Hiernaut's ones, we propose the following interpretation of the HBS behaviour during a thermal transient:

- Fission gas release due to HBS cracking induced by bubble over-pressurisation is not likely to exist, because it would lead to the cracking of HBS around all, in not all at least a lot of, bubbles, which is not observed.
- Burst releases are more likely to be due to the formation of cracks. These cracks would release the gas, that filled the bubbles opened by the crack formation. This interpretation is consistent with the sample microstructure change with the fracture of big samples observed by Hiernaut at 1,500K by SEM (see Figure 3). In this interpretation cracking is only depending on the sample geometry and its stress state.

Some effort is still needed in order to determine if the gas filling the bubbles of the HBS can be released even though they are reached by a crack, and hence by which mechanism. The study of the pre-existing crack opening during heat treatment is under way for that purpose.

CONCLUSIONS

The results of the annealing test coupled with ESEM study of a HBS sample from a high burn-up fuel pellet did not evidence significant mechanical degradation, contrarily to the results reported by Hiernaut, the microstructure is only affected by an opening of pre-existing cracks and micro cracks. We interpret this difference by considering that crack formation is responsible for burst release in Hiernaut's experiment.

ACKNOWLEDGMENTS

This work has been funded by the research program "Performance Burn-Up PERBU" run by M. Bauer at the CEA in collaboration with Electricité de France and Areva.

REFERENCES

1. Hj. Matzke and J. Spino, *Journal of Nuclear Materials* **248,** 170-179 (1997).
2. Hj. Matzke, *Journal of Nuclear Materials* **189,** 141-148 (1992).
3. H. Stehle, *Journal of Nuclear Materials* **153,** 3-15 (1988).
4. J. Spino et al., *Journal of Nuclear Materials* **231,** 179-190 (1996).
5. Y. Pontillon et al., *Proceedings of the 2004 Water Reactor Fuel Performance Meeting,* (Orlando, USA, 2004).
6. Y. Pontillon et al., *European Working Group "Hot Laboratories and Remote Handling",* *Plenary Meeting,* (Petten, the Netherlands, 2005).
7. Y. Pontillon et al., *2008 Water Reactor Fuel Performance Meeting,* (Seoul, Korea, 2008).
8. J.P. Hiernaut et al., *Journal of Nuclear Materials* **377,** 313-324 (2008).
9. M. Marcet et al, *Proceedings of Top Fuel 2009,* (Paris, France, 2009).

Mater. Res. Soc. Symp. Proc. Vol. 1215 © 2010 Materials Research Society — 1215-V16-45

Behaviour of nanocrystalline silicon carbide under low energy heavy ion irradiation

Dominique Gosset[1], Laurence Luneville[2], Gianguido Baldinozzi[3], David Simeone[1], Auregane Audren[4], Yann Leconte[4]

1: Matériaux Fonctionnels pour l'Energie, Équipe Mixte CEA - CNRS - École Centrale Paris, CEA Saclay, DMN SRMA LA2M, bat. 453, 91191 Gif-sur-Yvette, France
2: Matériaux Fonctionnels pour l'Energie, Équipe Mixte CEA - CNRS - École Centrale Paris, CEA Saclay, DM2S SERMA LLPR, bat. 470, 91191 Gif-sur-Yvette, France
3: Matériaux Fonctionnels pour l'Energie, Équipe Mixte CEA - CNRS - École Centrale Paris, CNRS, SPMS, Ecole Centrale Paris, Grande Voie des Vignes, 92295 Châtenay-Malabry, France
4: CEA Saclay, DSM IRAMIS SPAM, bat. 522, 91191 Gif-sur-Yvette, France

Abstract

Silicon carbide is one of the most studied materials for core components of the next generation of nuclear plants (Gen IV). In order to overcome its brittle properties, materials with nanometric grain size are considered. In spite of the growing interest for nano-structured materials, only few experiments deal with their behaviour under irradiation. To assess and predict their evolution under working conditions, it is important to characterize their microstructure and structure. To this purpose, we have studied microcrystalline and nanocrystalline samples before and after irradiation at room temperature with 4 MeV Au ions. In fact, it is well established that such irradiation conditions lead to amorphisation of the material, which can be restored after annealing at high temperature. We have performed isochronal annealings of both materials to point out the characteristics of the healing process and eventual differences related to the initial microstructure of the samples. To this purpose Grazing Incidence X-Ray Diffraction has been performed to determine the microstructure and structure parameters. We observe the amorphisation of both samples at similar doses but different annealing kinetics are observed. The amorphous nanocrystalline sample recovers its initial crystalline state at higher temperature than the microcrystalline one. This effect is clearly related to the initial microstructures of the materials. Therefore, the grain size appears as a key parameter for the structural stability and mechanical properties of this ceramic material under irradiation.

Introduction

In the frame of a sustainable energy production system, the Gen-IV project aims at the development of new nuclear reactors with better yields and safety as compared to the previous generations [1]. Different reactor concepts are considered, among which high temperature, fast neutron spectra reactors present interesting potentialities. However, some components of the core of such concepts require the use of materials able to withstand temperatures around 1000°C in normal conditions and 1300°C in incidental ones that can no longer be supported by the usual metal alloys. New materials such as ceramics are then to be used [2]. As such, silicon carbide has been considered as a possible structural component for high temperature fuel elements. Unfortunately, those materials are most often brittle. Reinforcement mechanisms are then to be considered aiming at making them acceptable. Two main routes are usually considered. First, fibre-composites can be elaborated in which cracks can hardly propagate as long as the fibres are not damaged. Second, nanostructured materials in which crack propagation is prevented by dispersion by the high grain boundaries density.

Reducing the grain size can have dramatic effects on the mechanical [3, 4] or transport properties [5] or the stability of the phases [6]. However, very few results exist regarding the specific effect of nanometric grain size on the behaviour of ceramics under irradiation.

Silicon carbide has for long been considered as diffusion barrier for fission products in the so-called Triso-fuel elements. More recently applications as possible material for structural elements of fuel components in some Gen-4 reactors concepts are proposed [7].

The structure of silicon carbide is based on the stacking of nearly regular, covalently bonded, Si-C4 tetrahedra, making close-packed hexagonal planes. Owing to the huge number of possible regular stackings of such planes with low energy differences, a large number of structures called polytypes can be observed [8]. A given polytype is named after the number of planes making the basic stacking and the initial of the resulting crystal structure. SiC can be classified into two generic families, β SiC, relatively stable at high temperature, has the only cubic structure of all polytypes the (3C), and most often contains a high density of irregularly distributed stacking faults, and α SiC, which is most often polyphasic, with its main components have hexagonal or rhombohedral structures (mainly 6H, 4H, 15R). Large monocrystals of the simplest polytypes with low stacking faults density can be obtained.

Irradiations of silicon carbide with neutrons, electrons, or slow heavy ions lead to two different behaviours, mainly depending on the irradiation temperature [9]. At low temperature (lower than 250 to 400°C, depending on the particle mass and energy), amorphisation is observed at low doses, low damage, around 0.3 dpa [10]. At high temperature, the material remains crystalline and the formation of extended defects (dislocation loops) is observed. The amorphous material can be healed by subsequent annealing. Two different processes have been identified. In the case of monocrystalline, partially damaged (e.g., subsequent to ion-doping) materials, epitaxial growth from the non-damaged bulk takes place [11]. This usually leads to the formation of crystalline material of the same polytype as the bulk, but with higher density of faults. This process is observed to occur at temperatures as low as 600°C. On the other hand, totally amorphous materials (e.g. after neutron irradiation) recrystallise by a classical nucleation and growth mechanism [12]: germs nucleate on pre-existing defects such as free surfaces. In that case, a very heterogeneous structure is observed, most often made of faulted 3C sub-micronic crystals with no epitaxial relationship with the initial material. This process is observed at high temperature, above 800°C.

In this paper, we compare the effect of irradiation by slow heavy ions and subsequent isochronal annealings of two kinds of samples: nanocrystalline β-SiC and micrometric, α. The analysis focuses on the influence of the different microstructures on the observed behaviours. The experiments are performed with X-ray diffraction in grazing incidence conditions (GIXRD) to analyse the structural (amorphisation, polytypes) and microstructural (microstrain, domains size) modifications occurring in the damaged layer of the materials.

Experimental

Materials

Nanometric β-SiC has been obtained via a classical laser-pyrolysis route [13]. A mixture of high purity silane and acetylene as Si and C precursors diluted in helium is used. The nanopowders have a mean grain size of about 20 nm and they exhibit a narrow size distribution. The powder batch was subsequently compacted by successive grinding and Cold Isostatic Pressing (CIP). The CIPed pellets are then high-isostatic pressure (HIP) hot-pressed at 1900°C under 200 MPa during 1h. The pellets have a relative density of 92%. They were then cut (discs diameter 20mm) and polished. Grain growth during HIP process was limited, leading to final grain sizes mainly in the 30 – 50 nm range, as determined from TEM observations. X-ray diffraction shows a cubic 3C polytype with line distortions usually attributed to a high density of stacking faults [14], as confirmed by TEM observations (Figure 1).

The micrometric α-SiC has been obtained by isostatic hot-pressing (HIP) of a commercial powder (H.C. Starck BF 17). The average grain size of this batch of powder is mainly in the submicron range (500 – 900 nm). The final density is about 90%. SEM observations show a limited grain growth (mean size around 1μm). Rietveld refinements of the XRD pattern show a polyphasic SiC material, mainly consisting of hexagonal 6H polytype coexisting with small proportion of rhombohedral 15R (~7 vol.%), hexagonal 4H (~4 vol.%) and residual graphite phases.

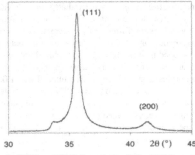 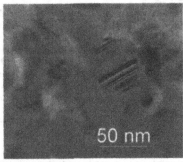

Figure 1: XRD and TEM analysis of the β-SiC material. The high faults density is responsible of the shoulder on the left of the (111) line

Irradiations

The irradiation conditions have been determined in order to simulate a neutron irradiation, i.e. most of the damage arises from ballistic collisions [15]. Here, 4 MeV Au ions have been used. The damage and ion range have been estimated with SRIM [16] (Figure 2). Taking into account recent evaluations of SRIM performances [17], the irradiations lead to the following estimations: damage mainly located from the surface to the implantation area, ion range around 840nm. From SRIM full cascades calculations, damages around 0.13 dpa at $1.10^{14}cm^{-2}$ and 0.66 dpa at $5.10^{14}cm^{-2}$ are estimated. In the damaged area, the nuclear energy loss and the electronic energy loss are both around 3 keV/nm. However, it has been shown the damage threshold for the electronic interactions in SiC is higher than 15 keV/nm [18]. It can then be assumed that most of damage is here produced by ballistic collisions.

X-ray diffraction

X-ray diffraction has shown to be a very efficient tool to analyse structural and microstructural modifications of materials under irradiation [19]. Here, the probed volume must be restricted to the damaged region of about 600 nm. An asymmetric setting with grazing incidence configuration (GIXRD) has then to be used [20]. In that case, the analysed depth directly depends on the incident beam angle and can be tuned to match the depth of the damaged region.

The incident beam is here a pure Cukα₁. A parallel beam is produced by a flat Ge (200) monochromator. A selection slit (50 μm) ensures the footprint of the X-ray beam on the sample is always smaller than the sample size even at the lowest grazing angle we used. The diagrams are obtained using a curved position-sensitive detector INEL CPS120. The analysis has been performed at a 0.4° incidence angle. In that case, nearly 80% of the analysed material is the damaged area, around 10% a mixed implanted and damaged slab, the remaining 10% the non-damaged bulk (Figure 2).

From the analysis of GIXRD diagrams, structural parameters (structural polytypes, main polytype, lattice parameters) and microstructural (crystalline vs amorphous fraction, size of the coherent diffraction domains, microstrains) can be deduced. Corrections specific to the grazing incidence setup must be taken into account (angle shifts, instrumental broadening of the peaks and intensity, we have introduced in a specific Rietveld program, XND [21]). Here, the lines are too broad for determining an accurate evolution of the cell parameters. Moreover, both materials appear highly textured. Therefore, we have focused the analyses to the identification of the crystalline phases identification and to the Hall-Williamson analysis of the peak widths.

Results and interpretation

Irradiations

The two materials, nanometric β-SiC and micrometric α-SiC were irradiated on the IRAMIS accelerator (JANNUS-Orsay facility [22]) at room temperature at two fluences, 1.10^{14} and $5.10^{14}/\text{cm}^2$ with a flux $3.10^{11}/\text{cm}^2.\text{s}$. After irradiation, GIXRD analysis was performed. The diffraction patterns are reported on Figure 4. Both materials exhibit the same evolution: at $1.10^{14}/\text{cm}^2$, a strong decrease of the intensity of the lines is observed. At $5.10^{14}/\text{cm}^2$, the damaged material is totally amorphised: the remaining diffraction lines correspond to the pristine bulk material. The corresponding damage is in good agreement with the previously established threshold for total amorphisation at room temperature, of about 0.3 dpa. The Hall-Williamson analysis of the widths of the lines was performed on the $1.10^{14}/\text{cm}^2$ diagrams (after removal of the instrumental width and restricted to non-overlapping main peaks: Figure 3). Relative to the unirradiated samples, this shows a strong broadening that can be attributed to high micro-strains. In the α-SiC microcrystalline sample a strong reduction of the size of the coherent diffraction domains is also responsible for a part of the observed broadening.

Though two materials with different initial structures, namely α and β SiC, and different microstructures, respectively nanometric and micrometric were irradiated, the results observed after irradiation can be explained by the only microstructural changes. First of all, the amorphisation in SiC irradiated at low temperature has been attributed to local disorganisation of the regular Si-C4 tetrahedra network [23]: no long-range defect diffusion is expected in such conditions. Therefore, the differences introduced by the initial polytypes are not significant and the measured initial grain size in the nanocrystalline sample is also too large to have any significant influence. This explains why the amorphisation of both materials occurs at the same fluence.

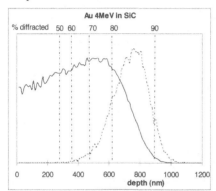

Figure 2: Au 4 MeV implantation of SiC (from SRIM, as re-evaluated by [17]). Solid line: damage (collisions). Doted line: implantation. % diffracted: fraction of material analysed in GIXRD conditions (Cukα₁ line, 0.4° incidence)

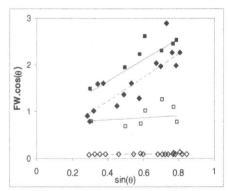

Figure 3: Hall-Williamson plots of the GIXRD diagrams. Empty symbols: non-irradiated. Full symbols: irradiated Au 4MeV $1.10^{14}/\text{cm}^2$; Lozenge: α-SiC. Squares: β-SiC. Lines: Hall-Williamson analysis.

166

Annealings

The most irradiated samples of both materials underwent isochronal annealings (from 800°C to 1100°C, 100°C steps of 30 minutes). This treatment allows the determination of the relative stability of the damaged materials and it provides an estimate of the temperature range for the activation of healing mechanisms. The X-ray diffraction diagrams we have obtained after each heating step show very different results (Figure 5). The α-SiC sample appears correctly recrystallised after the first heating at 800°C (diagram coherent with the 6H polytype) and shows nearly no modification up to 1100°C. On the contrary, the β-SiC sample has to be heated up to 1000°C before a significant healing is observed. After the annealings, the Hall-Williamson analyses show both materials have small coherent diffraction domains (both about 15 nm) and high residual microstrains.

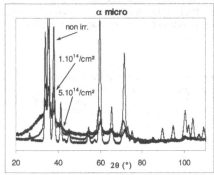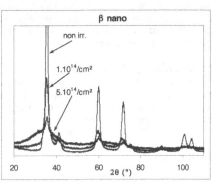

Figure 4: GIXRD diagrams for the α, micrometric and the β, nonametric SiC samples as a function of Au-4MeV fluence

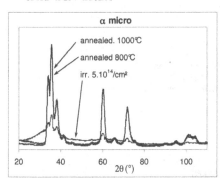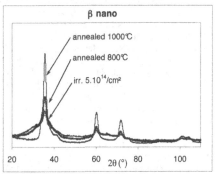

Figure 5: GIXRD diagrams obtained after isochronal annealings.

These results can be interpreted by considering the two known recrystallisation processes indentified in SiC. In the case of the α-SiC sample, it is worth noting that the grain size is larger than the thickness of the damaged material. In that case, epitaxial rebuilding can take place in most of the grains. This is known to occur at temperatures as low as 600°C but it generally leads to highly defective materials: this is coherent with both the diffraction patterns showing the formation of the 6H polytype and the Hall-Williamson analysis showing small, disordered, diffraction domains. On the other hand, healing of the β-

SiC sample is obtained in two steps: at the beginning, only the rebuilding of the grains located at the frontier between the damaged and bulk materials can occur by the same epitaxial growth process. But in the totally amorphised grains located in the damaged material, the second, nucleation and growth process has to be considered. In that case, faulted nanocrystals with a 3C structure are obtained because this is the most stable polytype in this temperature range.

Conclusion

Two different silicon carbide materials, respectively nanometric β-SiC and micrometric α-SiC were irradiated with Au 4 MeV ions, before undergoing isochronal annealings. Though these materials have different structures (polytypes) and microstructures, we show that the observed behaviours can be interpreted by their only microstructural features. They both show amorphisation at a fluence corresponding to the admitted amorphisation threshold. Healing the amorphous materials occurs via two different mechanisms, topotactic rebuilding for partially amorphised grains and nucleation and growth in the totally amorphised grains.

Acknowledgements

We wish to thank Lionel Thomé for performing the irradiations at the IRAMIS facility (JANNUS-Orsay) and the CEA-DEN/SRMA/LTMEX group for the preparation of the hot-pressed pellets. We are also very indebted to Isabelle Monnet (DSM IRAMIS CIMAP, GANIL, Caen) for performing the TEM observations. This work was supported by the French CEA-CNRS ISMIR CPR Research Program and the French National Research Agency project ANR-BLAN 56 ALIX MAI.

References

1 http://www.gen-4.org
2 P. Yvon, F. Carré, J. Nucl. Mat. **385** (2009) 217–222
3 I.E. Reimanis, Mat. Sci. Eng. **A237** (1997) 159-167
4 B. Jiang et al., Int. J. Plast. **20** (2004) 2007–2026
5 S. Surblé et al., Ionics **14** (2008) 33–36
6 G. Baldinozzi et al., Phys. Rev. **B 90-21** (2003) 216103
7 L.L. Snead et al., J. Nucl. Mat. **371** (2007) 329–377
8 R.W.G. Wyckoff, Crystal Structures, Vol. I. New York, Interscience (1963)
9 S.J. Zinkle, L.L. Snead, Nucl. Instr. Meth. Phys. Rev. **B-116** (1996) 92-101
10 W. Jiang, Y. Zhang, and W. J. Weber, Phys. Rev. **B70**, 165208 (2004)
11 M. Satoh et al., J. Appl. Phys. **89-3** (2001) 1986
12 L.L. Snead et al., Nucl. Instr. Meth. Phys. Res. **B-141** (1998) 123-132
13 N. Herlin-Boime et al., J. Nanop. Res. **6** (2004) 63-70
14 V.V. Pujar, J.D. Cawley, J. Am. Cer. Soc. **84[11]** (2001) 2645
15 L. Luneville et al., this meeting
16 J. Ziegler, http://www.srim.org
17 Yanwen Zhang et al., J. Appl. Phys., **105**, 104901 (2009)
18 A. Benyagoub et al., Appl. Phys. Lett. **89** (2006) 241914
19 D. Simeone et al., this meeting
20 D. Simeone, D. Gosset, CEA report CEA-R-5975 (2001)
21 J.F. Bérar, G. Baldinozzi, IUCR-CPD-Newletters **20** (1998)
22 Y. Serruys et al., Nucl. Instr. Meth. Phys. Res. **B-240** (2005), 124-127
23 W. Bolse, Nucl. Instr. Meth. Phys. Res. **B-148** (1999) 83-92

Metallic Materials II

Mater. Res. Soc. Symp. Proc. Vol. 1215 © 2010 Materials Research Society   1215-V17-04

Effect of Oversized Alloying Elements on Damage Rates and Recovery in Zirconium

Valeriy Borysenko, Yuri Petrusenko and Dmitro Barankov

CYCLOTRON Science & Research Establishment, National Science Center - Kharkov Institute of Physics & Technology, Kharkov, Ukraine

ABSTRACT

Studies were made into the influence of oversized rare-earth atoms on the processes of radiation defect accumulation and annealing in two-component zirconium alloys. Zr and Zr-X alloys (where X = Sc, Dy, Y, Gd and La) have been irradiated with 2 MeV electrons at 82 K. The radiation-induced resistivity has been measured in situ as a function of dose. As compared to unalloyed zirconium, the alloys have exhibited a decrease in the resistivity gain, this decrease being proportional to both the concentration and the size of dopant atoms. A possible explanation for the effect is offered. The difference between the recovery processes in zirconium and in its alloys has been studied. To this end, the irradiated specimens were subjected to isochronal annealing at temperatures between 82 and 350 K. It is shown that Dy, Y, Gd and La atoms trap interstitial atoms at stage I of the recovery. The dissociation of interstitial-impurity complexes takes place at stage II. In zirconium alloys with Dy, Y and Gd, splitting of recovery stage III into two substages has been revealed. The Zr-La alloy has not shown this splitting. Isothermal annealing data were used to determine the activation energies of recovery stages, and also to calculate the activation energy spectra for zirconium and its alloys. The oversized atoms of rare-earth metals are shown to interact effectively with both the interstitials and the vacancies in the zirconium matrix. This effect must be taken into account when developing new radiation-resistant Zr-base alloys or modifying the ones already existing.

INTRODUCTION

Currently the central problems in the radiation physics of metals and alloys are the following questions: investigations of properties of point defects and their interaction with alloying additions; radiation-stimulated segregation and phase transformations on irradiation in alloys; vacancy swelling and the behavior of transmutation gas impurities. By the present time much work has been done to investigated both theoretically and experimentally the point defect properties in a wide variety of metals. However, the processes of point defect interaction with impurity atoms, particularly in bcc and hcp lattices, are still not clearly understood.

Of special interest is the investigation into the interaction of radiation-induced point defects with substitutional atoms in the hcp lattice of zirconium. This paper reports the results from studies into the influence of oversized atoms of rare-earth metals on the kinetics of radiation-defect accumulation and annealing in the zirconium matrix.

EXPERIMENTAL DETAILS

Zirconium alloys with rare-earths (Sc, Dy, Y, Gd and La) have been prepared for the experiments. The alloys were melt in a laboratory arc furnace in a purified helium atmosphere. As a result, Zr-Sc, Zr-Dy, Zr-Y, Zr-Gd and Zr-La alloys with alloying element concentration of

0.15±0.02 wt.% were obtained. The alloys were used to make the samples. All the samples were annealed at 700° C under high vacuum (10^{-4} Pa) for 1 hour. The impurity concentration of zirconium and zirconium alloys is given in table I.

The resistivity of samples was measured by a standard four-probe method. A two-channel nanovoltmeter Agilent 34420A and two nanovoltmeters Solartron 7071 were used in the measuring system. All devices including the switches of samples, measuring current directions were connected by means of the IEEE-488 bus into a PC-controlled integrated system. This system enabled simultaneous measurements of both the resistivity and the temperature of each sample. In the experiment, 50 to 60 resistivity measurements of each sample were carried out in the temperature range from 78 to 81 K. The obtained results were approximated by the linear dependence, from which the resistivity value at 79.5 K was determined. The uncertainty of resistivity measurements was ± 5×10^{-12} Ωcm.

Table I. Concentration of impurity atoms in zirconium and its alloys.

Impurity	Impurity content (at. ppm)	Impurity	Impurity content (at. ppm)	Impurity	Impurity content (at. ppm)
Nb	<10	Mn	0.72	N	8.2
Hf	200	Pb	6.3	F	<5
Si	0.95	Fe	140	Mo	8.3
Al	1.3	Cr	35	Cd; Ca; Ti; K; Cl	<1
Ni	90	O	1100	B; Be; Li	<0.1
Cu	17	C	190	Zr	Balance

Irradiation was performed at the National Science Center – Kharkov Institute of Physics & Technology electrostatic electron accelerator ELIAS. The electron energy on the sample surface was 2 MeV, and the electron flux density was 10 μA/cm². The electron flux density was homogeneous within about ± 5 % across the irradiated area. Two identical sets of samples were exposed separately to an electron fluence of $1.4 \cdot 10^{19}$ e⁻/cm². The sample temperature during irradiation was about 82 K. The first set of irradiated samples was subjected to isochronal annealing according to the following scheme: in the temperature range from 83 to150 K with a temperature step of 2.4 K the annealing time was 6 min; and in the temperature range from 150 to 350 K with a 6.4 K temperature step the annealing time was 12 min. The second set of irradiated samples was subjected to a series of isothermal annealings.

EXPERIMENTAL RESULTS AND DISCUSSION

Figure 1 shows the increase of resistivity as a function of electron fluence for Zr and Zr-alloys. The dose dependences were used to calculate the initial damage rates $((d\rho/d\Phi t)_{\Delta\rho=0})$ for zirconium and its alloys. The cross-section for atomic displacements σ_d was calculated from the relation $(d\rho/d\Phi t)_{\Delta\rho=0} = \rho_F \sigma_d$, where ρ_F is the resistivity of a unit concentration of Frenkel pairs ($\rho_F = 37.5 \cdot 10^{-4}$ Ωcm [1]). The resulting $(d\rho/d\Phi t)_{\Delta\rho=0}$ and σ_d values for zirconium and its alloys are presented in table II along with the dopant concentrations and their radii [2].

172

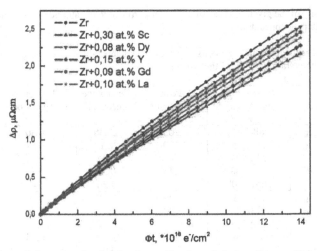

Figure 1. Increase of resistivity $\Delta \rho$ as a function of electron fluence Φt for Zr and Zr-alloys.

Table II. Atomic radii of dopants, radiation-induced resistivity gains, initial damage rates and damage cross sections for zirconium and zirconium alloys.

Sample	Atomic radii of the elements[2], $(10^{-1}$ nm)	Radiation-induced resistivity, $(\mu\Omega$ cm)	Initial damage rates $(d\Delta\rho/d\Phi t)_{\Delta\rho=0}$, $(10^{-25} \Omega$ cm^3/e$^-)$	Damage cross section σ_d, $(10^{-24}$ cm$^2)$
Zr	1.602	2.61621	2.216	59.1
Zr+0.30at.%Sc	1.641	2.14088	1.752	46.7
Zr+0.08at.%Dy	1.773	2.49426	2.110	56.3
Zr+0.15at.%Y	1.801	2.24065	1.889	50.4
Zr+0.09at.%Gd	1.802	2.42492	1.974	52.6
Zr+0.10at.%La	1.877	2.34289	1.928	51.4

It follows from the data that the introduction of oversized atoms of alloying elements into the zirconium matrix leads to deceleration of radiation-induced resistivity gain as compared with pure zirconium, and also, to a decrease in the initial damage rates and the damage cross section. This decrease is proportional to both the concentration and the size of dopant atoms. This effect can be explained from the following consideration. Since the chosen temperature of irradiation is close to the temperature of onset of free migration of interstitial atoms in zirconium [3-6], the observed resistivity gain retardation in alloys can be explained by additional interstitial-vacancy annihilation in the process of irradiation. This annihilation may be due to a decrease in the energy of interstitial migration in elastic strain fields created by oversized dopant atoms. It is reasonable to suggest that large-size dopant atoms give rise to large deformations in the lattice.

173

Therefore, the probability of the given process should be proportional to the atomic radius value and, obviously, to the dopant concentration.

Table III. Recovery stages, temperature intervals of stages, peak temperatures, percentage recovery at stages and activation energies of the processes

Sample	Stage	Temperature range of stages (K)	Peak temperature (K)	Percentage recovery	Activation energy of peak (eV)
Zr	I_E	88-113	105	28.36	0.30
	I_F	113-140	120	26.18	0.34
	II	140-205	-	9.54	-
	III	205-308	245	28.47	0.74
	-	above 308	-	7.26	-
Zr+0.30at. %Sc	I_E	88-113	105	27.21	0.30
	I_F	113-140	120	28.51	0.36
	II	140-205	-	9.32	-
	III	205-308	245	29.06	0.74
	-	above 308	-	5.91	-
Zr+0.08at.%Dy	I_E	88-113	105	25.18	0.29
	I_F	113-140	120	23.04	0.37
	II	140-205	156	13.00	0.46
	III	205-308	234 and 257	25.34	0.67; 0.79
	-	above 308	-	13.00	-
Zr+0.15at.%Y	I_E	88-113	105	25.74	0.30
	I_F	113-140	120	22.97	0.36
	II	140-205	159	15.59	0.47
	III	205-308	226 and 261	23.57	0.67; 0.80
	-	above 308	-	12.13	-
Zr+0.09at.%Gd	I_E	88-113	105	27.88	0.29
	I_F	113-140	120	22.54	0.35
	II	140-205	159	12.81	0.48
	III	205-308	228 and 268	23.96	0.69; 0.82
	-	above 308	-	13.59	-
Zr+0.10at.%La	I_E	88-113	105	26.61	0.30
	I_F	113-140	120	24.36	0.35
	II	140-205	154	13.48	0.44
	III	205-308	241	23.67	0.74
	-	above 308	-	11.61	-

Table III gives the numerical parameters of the recovery of irradiated zirconium and its alloys. Figure 2 shows the differential isochronal annealing curves of the materials. The peaks on the curves at temperatures of 105 K and 120 K correspond to the substages I_E and I_F of zirconium recovery [3]. At the substage I_E the free-migrating interstitial atoms annihilate with their own vacancy - this process is called correlated recombination. At the substage I_F the free-migrating interstitial atoms annihilate with other vacancies – this process is called uncorrelated recombination [3-7]. Gadolinium, dysprosium, lanthanum, or yttrium doping in zirconium result in a reduced recovery at substage I_E. As mentioned above, this reduction can be attributed to an additional annihilation of interstitial atoms with vacancies occurring in the process of irradiation. The given annihilation takes place as the interstitial atoms appear in the zone of action of elastic strain fields created by oversized atoms of dopants. This process tends to decrease the number of close Frenkel pairs responsible for the correlated recombination. On the other hand, the reduced recovery can be associated with the trapping of free-migrating interstitial atoms by dopant atoms in the process of annealing. The alloys under discussion exhibit a reduced recovery at substage I_F, too. At this substage, freely migrating interstitial atoms may be trapped by dopant atoms with the result that the recovery value will be decreased similarly to the substage I_E case. The detrapping of interstitial atoms occurs at stage II. This is evidenced by the increased recovery and by the appearance of new substages in the alloys in the temperature range of stage II. The stage II associated with rearrangement of interstitial clusters formed in the stage I and detrapping of interstitial atoms from impurities [7].

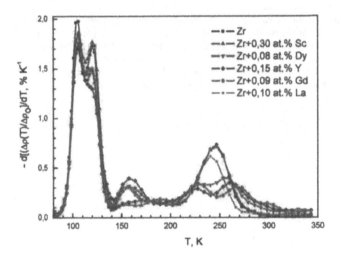

Figure 2. Differential curves of isochronal annealing of zirconium and its alloys irradiated with 2 MeV electrons at a temperature of 82 K to a fluence of $1.4 \cdot 10^{19}$ e⁻/cm².

For the Zr-Sc alloy, the recovery parameters are very close to similar parameters of pure zirconium. This is evidently due to the atomic radius similarity of the alloying element and the matrix and, as a consequence, to the absence of a noticeable interaction of scandium atoms with interstitials in the zirconium matrix.

The trapping of interstitial atoms by dopant atoms in zirconium alloys (except the Zr-Sc alloy) comes into conflict with the above-given suggestion about the presence of elastic repulsion forces between the atoms of oversized impurity and interstitials. This conflict can be resolved when the behavior of the nearest crystal lattice atoms that surround the defect or the

175

dopant atom is considered. The introduction of an oversized atom into the crystal lattice causes the lattice distortion. As a result, the atoms of the nearest shell are displaced in the direction from the dopant atom. The second shell of atoms shifts towards the dopant atom, and the third shell shifts in the direction from the dopant atom, etc. The value of the displacements decreases in inverse proportion to the distance from the center of the dopant atom to the third power. A similar pattern of displacements of the nearest atoms is also observed around the interstitial atom. Around the vacancy, similar displacements can be observed, but in the reverse order, i.e., the first layer of atoms is displaced towards the vacancy, the second layer is displaced away it, etc. [8, 9]. Thus, the field of displacements is strongly anisotropic, i.e., in different directions the displacement is of different sign and of different value. As a result, around the dopant atom there exist the regions of compression and extension. For this reason, in some directions the dopant atom repels the interstitial atom, while in the other directions it can attract the interstitial atom. It is reasonable to assume that around the oversized dopant atom in the extension regions there exist the positions, where the interstitial atom can be held till the beginning of stage II of the recovery. These positions should be found in the region of the second coordination sphere, where the greatest extension takes place.

It has been found that alloying of zirconium with dysprosium, yttrium or gadolinium causes stage III to split into two substages. However, in alloys with lanthanum or scandium the splitting is not observed. The stage III is associated with vacancy migrations [7]. The percentage of radiation defects in Zr-Dy, Zr-Y, Zr-Gd and Zr-La alloys, which remain after annealing at temperature of 308 K, is substantially higher than in unalloyed zirconium. This fact points to the formation of vacancy-impurity complexes in these alloys. The formation of similar complexes in the Zr-Sc alloy is not observed.

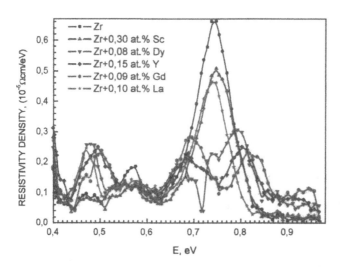

Figure 3. Activation energy spectra of recovery processes of zirconium and its alloys for stages II and III.

The isothermal annealing data were used to calculate the activation energies of the processes of the main recovery stages. To that end, the change-of-slope method was used [10]. The resulting activation energies are presented in table III. The activation energy spectra were calculated by the method of analysis proposed by W. Primak [11, 12]. In our case, to calculate the spectra at stages II and III we have used the second-order kinetic equation of recovery processes and then calculated in accordance with the procedure proposed in ref. [13].

The results obtained are shown in figure 3. The given spectra are coincident in shape with the isochronal annealing spectra (see figure 2). In the case of zirconium alloying with yttrium or gadolinium, the energy range corresponding to the detrapping of interstitial atoms from dopant traps is wider (0.44 – 0.54 eV) in comparison with that of Zr-Dy and Zr-La alloys (0.44 – 0.50 eV). This may imply that for yttrium and gadolinium atoms as impurity traps there exist, at least, two different configurations of interstitial atoms with different binding energies. It should be noted that with an increasing size of the alloying atom there occurs extension in the binding energy range of interstitial and dopant atoms. This range also has the trend to shift to higher binding energy values. However, this regularity is violated for the Zr-La alloy, which shows the lowest energy range. In this case, of all the elements studied here, lanthanum has the largest atomic radius. As demonstrated above, for the Zr-La alloy there is also no splitting of stage III into two substages, which falls out of the frame of the tendency for splitting observed as the size of the dopant atom increases. To explain the present experimental data on the of recovery processes at stage III, further investigations are required.

CONCLUSIONS

The present experimental data and their analysis suggest the following conclusions.
1. The introduction of oversized atoms of Sc, Dy, Y, Gd or La into a zirconium matrix retards the growth in resistivity dependent on the electron fluence, and also decreases the initial damage rates and the damage cross section in the process of irradiation with 2 MeV electrons at the temperature 82 K. The decrease correlates with the concentration and atomic radius of alloying elements.
2. Around oversized atoms of Sc, Dy, Y, Gd or La in the zirconium matrix there exist the regions of compression and extension.
3. Dy, Y, Gd and La atoms trap interstitial atoms at substage I_F of recovery. The interstitials are retained in the positions located in the region of the second coordination sphere. The detrapping of interstitials takes place at stage II of the recovery.
4. It has been found that alloying of zirconium with dysprosium, yttrium or gadolinium leads to splitting of stage III into two substages. However, this effect is not observed for the Zr-La and Zr-Sc alloys.
5. Vacancy-impurity complexes are formed in the Zr-Dy, Zr-Y, Zr-Gd and Zr-La alloys. The Zr-Sc alloy has shown no signs of vacancy-impurity complex formation.
6. With an increasing size of the alloying atom there occurs extension of the binding energy range of interstitial and dopant atoms towards higher energies. This regularity is violated for the Zr-La alloy, which shows the lowest energy range.

ACKNOWLEDGEMENTS

The authors express their gratitude to P.N. V'yugov for zirconium alloys kindly provided for our studies.

REFERENCES

1. C. H. M. Broeders and A. Yu. Konobeyev, *J. Nucl. Mater.* **328**, 197 (2004).
2. W.B.Pearson. *The Crystal Chemistry and Physics of Metals and Alloys* (Wiley Interscience, New York, 1972) Ch.4.
3. H. H. Neely, *Can. J. Phys.* **46**, 321 (1968).
4. H. H. Neely, *Rad. Eff.* **3**, 189 (1970).
5. M. Biget, F. Maury, P. Vajda, A. Lucasson and P. Lucasson, *Rad. Eff.* **7**, 223 (1971).
6. F. Dworschak, C. Dimitrov and O. Dimitrov, *J. Nucl. Mater.* **82**, 148 (1979).
7. W. Schilling and K. Sonnenberg, *J. Phys. F : Metal Phys.* **3**, 322 (1973).
8. A. Scholz and C. Lehmann, *Phys. Rev.* **B6**, 813 (1972).
9. C. N. Tome, A. M. Monti and E. J. Savino, *Phys. Stat. Sol.* **B92**, 323 (1979).
10. A.C.Damask and G.J.Dienes, *Point defects in metals*, (1963), p.146.
11. W. Primak, *Phys. Rev.* **100**, 1677 (1955).
12. W. Primak, *J. Appl. Phys.* **31**, 1524 (1960).
13. F. Dworschak, K. Herschbach and J. S. Koehler, *Phys. Rev.* **133**, A293 (1964).

Complex Materials and Devices

Mater. Res. Soc. Symp. Proc. Vol. 1215 © 2010 Materials Research Society   1215-V18-01

Materials Development by a Surface Modification for the Sulfuric Acid Decomposer in Iodine-

Sulfur(IS) cycle for Nuclear Hydrogen Production System.

Jae-Won Park, Hyung-Jin Kim and Yongwan Kim

Korea Atomic Energy Research Institute, Daejon-City, South Korea

ABSTRACT
Efforts have been made to develop long term sustainable materials for use above 900°C in the SO3/SO2 environments utilized in Nuclear Hydrogen Production Systems. In this study, the surface of Hastelloy X was serially modified by evaporative SiC coating, ion beam mixing (IBM), additional coatings, and final ion beam hammering (IBH). Subsequent heating above 900°C results in no peeling-off of the SiC coating layer in spite of the huge difference in their coefficients of thermal expansion (CTE). It was also found that ion beam hammering (irradiation) suppresses the vacuum sublimation of the low density (~40% of bulk) ceramic film (bulk materials begin to sublime at ~ 750°C in a vacuum of 1.5×10^{-5} torr). The sublimation rate is $\geq 30\%$ of the bulk rate after an annealing at 950 for 2 hrs but is decreased to $\leq 10\%$ of the bulk rate after irradiation with 70 keV ions to a total dose of 1×10^{17} N ions/cm^2. Further irradiation (up to 4×10^{17} N ions/cm^2) does not further decrease the rate. Both an immersion test in 98% sulfuric acid and the potentiodynamic polarization test suggest that the surface modified Hastelloy X has a greatly prolonged life time in the corrosive sulfuric acid atmosphere, suggesting the serial surface modification process is applicable to the thermo-chemical system for the nuclear hydrogen production.

INTRODUCTION
Since it was reported that massive production of hydrogen could be possible using the high temperature reactor with the aid of thermochemical methods [1], High Temperature Gas Cooled Reactor (HTGR) combined with the Iodine-Sulfur (IS) cycle has been regarded as the most efficient system for a mass production of hydrogen [2-3]. In the IS cycle, a process heat exchanger (PHE) comprised of channels for He and decomposed sulfuric acid gas (SO2/SO3/H2O) is needed. The material used for the sulfuric acid gas channels is of importance because it is subjected to severe corrosion environment. Currently there is no suitable commercial metallic material available. To surmount this obstacle, consideration is being given to surface modification of metallic materials or a development of ceramic PHEs. In this work, the surface modification option was studied. We selected Hastelloy X as the metallic substrate due to its good mechanical properties at a high temperatures and SiC as a corrosion inhibiting coating material (its corrosion resistance being due to the very strong covalent bonding between silicon and carbon [4-5]). This ceramic-coating-on-metal system certainly has considerable merit because that it does not hamper the manufacturabilty of the system as compared to a ceramic PHE system. In developing a surface coated metallic system, consideration needs to be given to how best to ensure the adhesion of the coating layer and how to reduce the defects (such as micro-porosities and micro-pipes) in the coating. Another potential disadvantage of ceramic coated metallic substrates is thermally induced delamination (CTE of Hastelloy X:16.6x10-6 at 980°C and CTE of SiC :5.0x10-6 at 1000°C). However we have solved this problem by employing an ion beam mixing (IBM) [6]. Other potential problems include the possible sublimation of the deposited film during a high temperature annealing in vacuum (the evaporation or sublimation of a condensed phase rather than melting followed by the evaporation or sublimation tends to occur at a lower ambient

181

pressure and a higher temperature, than is suggested by Clausius-Clapeyron Equation). Under a certain condition, the density of the condensed matter is susceptible to the sublimation. This sublimation certainly gives arise to adverse effects such as loss of the film and/or its deterioration and non-stoichiometric structure formation [7].

In this paper, we show a review of the strong adhesion produced by IBM, a result of in-depth sublimation suppression study of the SiC coated layer, and the corrosion study results of an immersion test in a 98 % sulfuric acid at 300 and a room temperature Potentiodynamic polarization test.

EXPERIMENT

SiC films were deposited with an electron beam evaporative method on metallic substrate (Hastelloy X sheet: ~ 15x15x0.5mm) surface-polished by a diamond paste up to 0.5 µm. Prior to a SiC deposition, a sputter cleaning of the sample was carried out for 10 minutes with an N ion energy of ~10 keV and a current of 0.5 Ampere. Then, the electron beam evaporative deposition of the SiC was performed to 50 nm thickness, followed by a nitrogen ion beam mixing at 70keV with a dose of ~ $5x10^{16}$ ions/cm^2. The ion beam mixing was to ensure no peeling-off of the film during heating [7]. A further SiC evaporative deposition up to a total of ~ 1 µm was then conducted with a deposition rate of ~3 Å/s produced by an electron beam current of ~0.15 A. The substrate temperature during the e-beam evaporative deposition was ~150°C. It is the outermost surface of the samples that was again N-irradiated (hammered) at 70 keV with a dose of ~ $1x10^{17}$ ions /cm^2 to $4x10^{17}$ ions/cm^2 to increase their resistance to sublimation.

Although the thickness and growth rate of the thin films were detected by a gold plated quartz crystal, the thickness of the film deposited in a controlled way was finally confirmed with the cross sectional SEM observation. The samples were then placed in an alumina boat and annealed in a quartz tube vacuum furnace with a heating rate of ~7.5°C/min at temperature ranges of 550°C to 950°C for 2 hrs. The vacuum chamber was evacuated by means of mechanical vacuum pumping and then an oil diffusion pumping. The absolute vacuum degree reached ~1.5 $x10^{-5}$ torr by continuous operation of the pumping system. The metallic substrate before the film deposition and the coated samples were weighed before and after annealing with a micro balance with a readability of 0.01mg (a repeatability of 0.02mg) to determine the sublimation rate.

Immersion corrosion test was performed in 98% H_2SO_4 at 300°C during about 350hrs to see the difference between the bare and. the coated samples. Conventional electrochemical polarization cell was used for studying electrochemical behavior of SiC coated Hastelly X samples at room temperature in an aerated solution that atmospheric air was directly pumped into the solution. Sulfuric acid solutions with 1 N concentration were used in this study. The counter electrode was a platinum wire. Saturated calomel electrode (SCE) was applied as a reference electrode with respect to standard hydrogen electrode (SHE).

RESULTS AND DISCUSSION

The surface of the film is covered with thick oxide layers and the oxides are found through the film although the concentration is reduced as going deeper inside from the surface of the film. The oxides appear to be silicon oxides. As determined by the weight changes, the surface areas of the sample, and the SiC film thickness, the density of the SiC films in this analysis was about 1.92 g/cm^3. That is about 40% lower than that of the bulk SiC (3.217 g/cm^3) [8]. After heating at 950°C in both air and vacuum, SiC film was maintained as-coated on the Hastalloy-X surface in spite of a high difference in their CTEs. This is attributed to IBM to produce a highly adherent coated layer and an interfacial reaction during annealing [6]. The reason N ions were used in preference to other inert gas ions is that N-ion bombardment makes not only the interface mix but also the metallic substrate reinforce by forming nitrides by implantation into the substrate.

It turned out that the SiC film was not detached from the substrate at the elevated temperature, but the sublimation of the film occurred. Visually the surface of the as-deposited SiC film is yellowish and is found to be grayish after an annealing at 950°C. This color change should be due to the crystallization of the SiC. We also observed a coating on the alumina boat where the samples were located and on the inner wall of the quartz annealing tube. XPS analysis reveals that the coating contains Si and C elements, implying that the SiC film was transported to the alumina boat during an annealing [9]. The XPS also showed that the SiC film deposited on the Hastelloy X is still remained even after the sublimation occurred. As shown in Fig. 1, almost no sublimation occurs below 700≤, but the sublimation rate increases drastically as the temperature increases, that is, ~ 5% at 750°C, ~ 7% at 850°C, and ~ 30% at 950 (Fig 1). The sublimation rate was determined by the weight decrease of the film with a careful exclusion of the substrate effects.

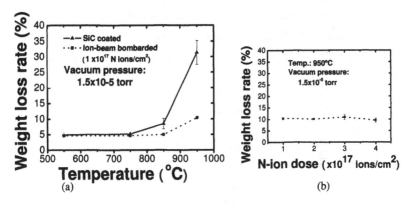

Figure 1. The density of an e-beam-evaporative deposited SiC film is about 60% of the bulk SiC. The sublimation occurs vigorously at 950°C, but it is greatly suppressed upon the N ion beam bombardment onto the film (a). At 70 keV, ion doses more than $1 \times 10^{17}/cm^2$ do not affect the sublimation rate that much (b).

According to the Clausius-Clapeyron equation, the sublimation rate should be even higher at the lower pressure and the higher temperature. However, when the N ions at 70 keV with a dose of 5×10^{17} ions/cm^2 were bombarded onto the film and then annealed in a vacuum of ~1.5×10^{-5} torr, no green color was observed on the alumina boat after an annealing at 950≤, implying that the sublimation was suppressed. However, the sublimation rate is still measured by ~ 8% at 950°C.although the ion beam irradiation was performed onto the film, and it is not decreased further as the ion dose increases up to 4×10^{17} ions /cm^2. The reason why the ion beam irradiation does not eliminate the sublimation completely may be that the nuclear collisions between the impinging ions and the atoms in the nearer surface of the deposited SiC film are less probable [9]. That is, the sublimation may have occurred on the surface until the most ion stopping range, the most densified layer, is exposed. Since the ion beam bombardment was performed at the normal incidence to a sample surface, the sputtering of the SiC film by the ion beam bombardment is very little (<1) [11]. We believe that the ion beam irradiation densifies the films due to a hammering effect [12]. The SiC film is fully crystallized after an annealing at 950≤. Since no weight reduction of the bulk SiC is found during this vacuum annealing, the vacuum sublimation of the SiC film is likely to occur mostly before it is crystallized [9].

Figure 2 shows weight retain rates of the as-received Hastelloy X and the both side coated Hastelloy X samples as a function of the immersion time (hours) in 98% sulfuric acid at 300°C. The coated sample with IBM retains the weight almost unchanged, but the weight of the uncoated reduces rapidly. The process heat exchanger is supposed to be used in the environment of the decomposed sulfuric acid gas above 900°C, but this test is performed in a 98% sulfuric acid at 300°C which is a very severe corrosion environment and exhibits the difference in a shorter period.

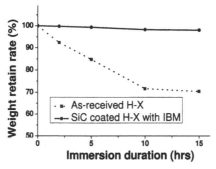

Figure 2. Corrosion rates of the as-received and SiC coated Hastelloy X samples in 98% sulfuric acid at 300°C.

Figure 3. is a cross sectional macroscopic observation of as-received and SiC coated Hastelloy X sheets bending-machined as a real shape for the decomposed sulfuric acid channel before and after immersing in 98% H_2SO_4 for 80 hrs and 348hrs at 300≤ . The upper two sheet samples are SiC coated samples annealed in vacuum and air at 950°C. The lowest is the as-received sample. The annealing atmosphere does not affect the corrosion behavior. Even after 80hrs immersion we can see a big difference in the thickness of the samples and, after 348 hrs, the as-received sample is corroded out from the edge. Although the corrosion rate is quite different, the thickness of the coated sample is also reduced. This may be attributed to the progress of the corrosion through the coalescent pores (micro-pipes) formed at the grain boundaries during the crystallization of the amorphous SiC film. The corrosion through the coalescent pores may have resulted in the film peeling-off. A repeated process of the coating and annealing is being conducted to reduce the existence of the micro-pipes in order to improve the corrosion property.

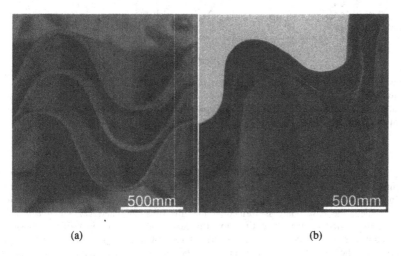

(a)   (b)

Figure 3. A cross sectional observation of as-received and the SiC coated Hastelloy X sheets shaped for the decomposed sulfuric acid channel before and after immersing in 98% H_2SO_4 for 80 hrs (a) and 348hrs (b) at 300°C.

Fig. 4 shows potentiodynamic polarization curves of the SiC coated Hastelloy X sample in aerated 1 N H_2SO_4 at room temperature. Corrosion potential (Ecorr) and corrosion current density(Icorr) of the non-coated are -240mV mV and ~6μA/cm^2, while those of the coated are -150mV mV and ~1 μA/cm^2. The coated shows a more noble corrosion behavior, but the difference is not so big as expected. It seems the Hastelloy X already has a high corrosion resistance in the sulfuric acid at room temperature. The potentiodynamic polarization analysis at the elevated temperature seems to be needed to compare with the weight reduction test in 98% sulfuric acid at 300°C. It is clear that Hastelloy X has a good corrosion resistance in sulfuric acid, but the SiC coating improves the corrosion resistant even further.

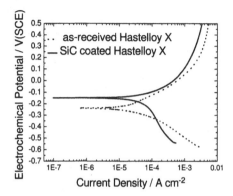

Current Density / A cm^{-2}

Figure 4. Potentiodynamic polarization curves of the SiC coated Hastelloy X sample in aerated 1 N H$_2$SO$_4$. shows that corrosion potential (Ecorr) and corrosion current density(icorr) of non-coated are -240mV mV and 6 μA/cm^2, while those of the coated are -150mV mV and 1 μA/cm^2.

CONCLUSIONS

We report the SiC coating on Hastelloy X can be sustained at a high temperature in spite of the hugh difference in CTE, when IBM is applied. The mechanism may be that the ion beam mixing fastens the SiC coated layer with the Hastelloy X substrate until the interfacial reaction occurs. Then, the new phase is formed at the film/substrate interface, which acts as a functionally degraded layer. The final ion irradiation on the deposited film produces a sublimation barrier. The corrosion tests of the resultant samples suggest that the SiC coated Hastelloy X should improve the lifetime of the process heat exchanger. The PHE system manufactured in this way certainly has a merit that it does not hamper the manufacturability of process heat exchanger of I-S cycle as compared to the whole ceramic system.

ACKNOWLEDGMENTS
This study was supported by both Nuclear Hydrogen Development Project and Proton Engineering Frontier Project sponsored by Ministry of Education, Science and Technology, Republic of Korea.

REFERENCES

[1]] S. Yalcin, International journal of hydrogen energy, Vol. 14, No8 (1989) 551

[2] Hiroyuki Ota, Shinji Kubo, Masatoshi Hodotsuka, Takanari Inatomi, Masahiko Kobayashi, Atuhiko Terada, Seiji Kasahara, Ryutaro Hino, Kenji Ogura, Shigeki Maruyama, 13th International Conference on Nuclear Engineering, Beijing, China, May 16-20, 2005, ICONE-13-50494

[3] S. Fujikawa et. al., J. Nucl. Sci. Technol., Vol. 41 (2004) 1245

[4] J.-P. Riviere, J. Delafond, P. Misaelides, F. Noli, Surf. Coat. Technol. 100-101 (1998) 243

[5] S. Fujikawa, H. Hatashi, T. Nakazawa, K. Kawasaki, T. Iyoku, S. Nakagawa and N. Sakaba., J. Nucl. Sci. Technol., 41 (2004) 1245

[6] J. Park, Z. S. Khan, H. Kim, Y. Kim, Mater. Res. Symp. Proc., Vol. 1125 (2009)65

[7] Richard N. Grugel a,, Houssam Toutanji, Advances in Space Research 41 (2008) 103

[8] CRC hand book of chemistry and physics,(CRC press Inc., 1988) B-60, ISBN 0-8493-0740-6

[9] J. Park, H. Kim, Y. Kim, Submitted for the publication else where (2009)

[10] J. F. Ziegler, J. P. Biersack, U. Littmark, "The Stopping Range of Ion in Solids", Pergamon Press, New York (1985)

[11] A. Zalar, J. Kovac, B. Pracek, P. Panjan, M. Ceh, Applied. Surface Science, Vol. 254, issue 20 (2008) 6611–6618

[12] B. Canut, V. Teodorescu, J.A. Roger, M.G. Blanchin, K. Daoudi, C. Sandu, Nuclear Instruments and Methods in Physics Research B, 191 (2002) 783–788

Mater. Res. Soc. Symp. Proc. Vol. 1215 © 2010 Materials Research Society   1215-V18-02

Thermal and Mechanical Properties of Hf Hydrides with Various Hydrogen Content

Ken Kurosaki[1], Masato Ito[1,*], Yuki Kitano[1], Hiroaki Muta[1], Masayoshi Uno[2], Kenji Konashi[3], and Shinsuke Yamanaka[1,2]

[1]Graduate School of Engineering, Osaka University, 2-1 Yamadaoka, Suita, Osaka 565-0871, Japan
[2]Research Institute of Nuclear Engineering, Fukui University, 3-9-1 Bunkyo, Fukui 910-8507, Japan
[3]Institute for Materials Research, Tohoku University, 2145-2, Narita, Oarai-machi, Ibaraki 311-1313, Japan
*Present address: Mitsubishi Materials Corporation

ABSTRACT

Fine bulk samples of delta-phase Hf hydride with various hydrogen contents (C_H) ranging from 1.62 to 1.72 in the atomic ratio (H/Hf) were prepared, and their thermal and mechanical properties were characterized. In the temperature range from room temperature to around 650 K, the heat capacity and thermal diffusivity of the samples were measured and the thermal conductivity was evacuated. The elastic modulus was calculated from the measured sound velocity. The Vickers hardness was measured at room temperature. Effects of C_H and/or temperature on the properties of Hf hydrides were discussed. At room temperature, the thermal conductivity values of the Hf hydrides were 23 $Wm^{-1}K^{-1}$. The Young's and shear moduli and the Vickers hardness of Hf hydride decreased with increasing C_H.

INTRODUCTION

Although boron carbide (B_4C) is currently expected as neutron control materials for fast reactors (FRs), the life time of the B_4C control rods is restricted by pellet-cladding mechanical interaction failure due to the helium gas swelling which is caused by the following nuclear reaction:

$$^{10}B(n,\alpha)^7Li.$$

In order to prolong the life time of the control rods, we have proposed Hf hydrides as the neutron control materials [1], because no helium gas is produced during neutron absorption. The nuclear reaction is expressed as follows:

$$^{177}Hf(n,\gamma)^{178}Hf(n,\gamma)^{179}Hf(n,\gamma)^{180}Hf.$$

Since hydrogen can moderate fast neutrons and Hf has large cross sections for thermal neutron capture, Hf hydrides would be an effective neutron control materials in FRs. In addition, since the daughter nuclides [178]Hf and [179]Hf also have large cross sections for thermal neutron capture as well as the parent nuclide [177]Hf, Hf hydrides would absorb neutrons for more than 40 years [1].

When utilizing Hf hydrides as long-life neutron control materials with no helium production in FRs, it is very important to understand their thermal and mechanical properties. Although the physical properties of Ti hydrides [2] and Zr hydrides [3] have been widely examined, those of Hf hydrides have been scarcely reported until now. Recently, the authors succeeded to synthesis fine bulk samples of Hf hydrides with various hydrogen contents and evaluated their thermal and mechanical properties at elevated temperatures up to around 650 K [4,5]. In the present paper, we will summarize our progress in understanding the effects of C_H and/or temperature on the thermal and mechanical properties of Hf hydrides.

EXPERIMENT

It is generally considered that fabrication of crack-free bulk metal-hydrides is quite difficult because there exists large volumetric change occurred due to phase transition from the metal phase to the hydride phase during hydrogenation. In the present study, we have succeeded to prepare various shapes of bulk Hf hydrides with no cracks and pores, by developing an appropriate method to prevent pulverization during hydrogenation as much as possible.

The hydrogen content (C_H = H/Hf) of the Hf hydride samples was checked both by a hydrogen analyzer and by mass change before and after the hydrogenation. The obtained hydride samples were characterized with an x-ray diffraction (XRD) method using Cu $K\alpha$ radiation at room temperature, and the lattice parameter was calculated from the obtained XRD pattern. The density of the hydride samples was calculated based on the measured weight and dimensions.

The heat capacity (C_P) was measured in an argon atmosphere with a flow rate of 100 ml/min using a differential scanning calorimeter from room temperature to approximately 673 K. The thermal diffusivity (α) was measured by the laser flash method from room temperature to approximately 650 K in vacuum ($\sim 10^{-4}$ Pa). The thermal conductivity (κ) was evaluated from the thermal diffusivity (α), heat capacity (Cp), and the sample density (d) based on the following relationship:

$$\kappa = \alpha \cdot C_p \cdot d .$$

The electrical resistivity (ρ) was also measured by the standard four-probe dc method in a helium atmosphere (~ 0.05 MPa) from room temperature to approximately 650 K.

The elastic modulus of Hf hydride was evaluated from the ultrasonic pulse echo method. The samples were bonded to a 5 MHz longitudinal or shear sound wave echogenic transducer and the longitudinal (V_L) and shear (V_S) sound velocities were measured. The measurements were performed at room temperature in air. From the measured V_L and V_S, the Young's (E) and shear (G) moduli were calculated based on the following equations:

$$E = \frac{dV_S^{\,2}\left(3V_L^{\,2} - 4V_S^{\,2}\right)}{\left(V_L^{\,2} - V_S^{\,2}\right)},$$

$$G = dV_S^{\,2}.$$

The Vickers hardness (H_V) was also measured using a Vickers harness tester at room temperature in air. The applied load and the load time for the indentation tests were set to 9.8 N and 10 sec, respectively. The indentation at the present experimental condition did not induce cracking in specimens. The H_V was calculated based on the following relationship:

$$H_V = \frac{2P_H \sin(\phi/2)}{l^2},$$

where ϕ is the indenter apex angle (in this case, 136°), P_H is the applied load, and l is the mean length of diagonal of the indentation trace. In the present study, the H_V was evaluated by dividing the load by the indentation area in GPa unit.

RESULTS AND DISCUSSION

We succeeded to obtain various shapes of bulk Hf hydride samples [4]. The C_H values of the prepared samples were in the range from 1.62 to 1.72 in the atomic ratio. From the XRD analysis, all the samples were identified as a cubic delta phase with the fluorite-type structure (fcc_C1). For example, the XRD pattern of the sample whose C_H is 1.63 is shown in Figure 1, together with the literature data of δHfH$_{1.64}$ [6]. The XRD pattern of our sample was well consistent with the literature data. Figure 2 shows the cubic lattice parameter at room temperature for the Hf hydrides as a function of the C_H, together with the literature data reported by Sidhu [6,7], Espagno [8], Katz [9], and Lewis [10]. The lattice parameter values of our samples were slightly larger than those of the literature values. It was revealed that the lattice parameter of the Hf hydrides linearly increased with increasing the C_H.

Figure 3 shows the temperature dependence of the heat capacity (C_P) of δHfH$_{1.64}$, δHfH$_{1.70}$, and pure Hf, together with the literature data of pure Hf reported by Cox [11]. It can be confirmed that our C_P data for pure Hf were completely consistent with the literature data, indicating that our C_P data were reliable. The C_P values of the Hf hydrides were higher than that of the pure Hf and slightly increased with the C_H especially at high temperatures. In addition, the C_P values of the Hf hydrides increased with temperature unlike that of pure Hf. This increase in the C_P with both temperature and the C_H would be attributable to excitation of the optical hydrogen vibration.

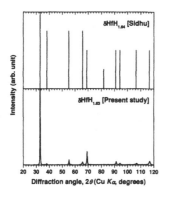

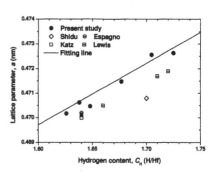

Figure 1. X-ray diffraction patterns of δHfH$_{1.63}$, together with the literature data of δHfH$_{1.64}$.

Figure 2. Lattice parameter at room temperature vs. hydrogen content for δHfH$_x$, together with the literature data.

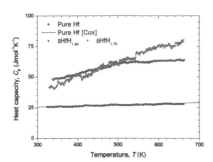

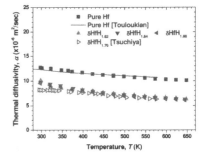

Figure 3. Temperature dependence of the specific heat capacity of pure Hf and δHfH$_x$, together with the literature data.

Figure 4. Temperature dependence of the thermal diffusivity of pure Hf and δHfH$_x$, together with the literature data.

Figure 4 shows the temperature dependence of the thermal diffusivity (α) of δHfH$_{1.62}$, δHfH$_{1.64}$, δHfH$_{1.68}$, and pure Hf, together with the literature data of δHfH$_{1.70}$ and pure Hf reported by Tsuchiya [12] and Toloukian [13], respectively. Similar to the C_P results, our α data for pure Hf were completely consistent with the literature data. It was revealed that the α values of the Hf hydrides were apparently lower than that of pure Hf, and slightly decreased with the C_H. A good agreement between our data and the literature data [12] for the Hf hydrides can be confirmed. By using these α values, C_P data, and measured density values, we calculated the thermal conductivity (κ) of the Hf hydrides.

Figure 5 shows the temperature dependence of the κ of $\delta HfH_{1.62}$, $\delta HfH_{1.64}$, $\delta HfH_{1.68}$, and pure Hf, obtained in the present study. At room temperature, the κ values of the Hf hydrides ($\sim$23 $Wm^{-1}K^{-1}$) were almost consistent with that of pure Hf, irrespective of C_H values. However, at temperatures higher than 400 K, the Hf hydrides indicated slightly large κ values compared to that of pure Hf, and the tendency was more remarkable in the samples with larger C_H.

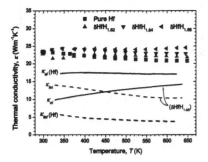 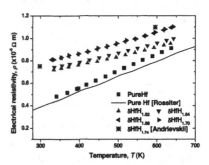

Figure 5. Temperature dependence of the thermal conductivity of pure Hf and δHfH_x; the electronic and lattice contributions to the thermal conductivity of pure Hf and $\delta HfH_{1.68}$ are also plotted.

Figure 6. Temperature dependence of the electrical resistivity of pure Hf and δHfH_x, together with the literature data.

Next, in order to discuss the magnitude relation of κ between Hf hydrides and pure Hf, the electrical resistivity (ρ) were evaluated because the total (measured) κ can be described as the sum of the lattice and the electronic contributions and the latter is closely related with the electrical conductivity (or resistivity) through the Wiedemann-Franz relation, $i.e.$:

$$\kappa_{total} = \kappa_{lat} + \kappa_{el} = \kappa_{lat} + \frac{LT}{\rho},$$

where κ_{total} is the total thermal conductivity, κ_{lat} is the lattice contribution to the thermal conductivity, κ_{el} is the electronic contribution to the thermal conductivity, and L is the Lorenz number (= 2.45×10^{-8} $W\Omega K^{-2}$). Figure 6 shows the temperature dependence of the ρ of $\delta HfH_{1.62}$,

193

$\delta HfH_{1.64}$, $\delta HfH_{1.68}$, $\delta HfH_{1.70}$, and pure Hf, together with the literature data of $\delta HfH_{1.74}$ and pure Hf reported by Andrievskii [14] and Rossiter [15], respectively. The ρ data of the Hf hydrides indicated metallic temperature dependence, *i.e.* $\Delta\rho/\Delta T > 0$. The ρ values of the Hf hydrides were larger than that of pure Hf and increased with the C_H. In both Hf hydrides and pure Hf, our ρ data were well consistent with the literature data [14,15]. The $\Delta\rho/\Delta T$ values decreased with increasing the C_H in the Hf hydrides. This tendency has been observed in the related delta-phase metal hydrides viz. δZrH_x [16] and δTiH_x [17].

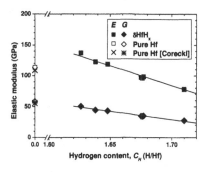

Figure 7. Young's (E) and shear (G) moduli of pure Hf and δHfH_x as a function of C_H; The reference value for pure Hf is also plotted for comparison.

Figure 8. Vickers hardness (H_V) of pure Hf and δHfH_x as a function of C_H.

By using the ρ data, we estimated the κ_{lat} and κ_{el} for $\delta HfH_{1.68}$ and pure Hf, and plotted the results in Figure 5 as a function of temperature. The κ_{el} of $\delta HfH_{1.68}$ gradually increased with temperature, while that of pure Hf indicated rather flat temperature dependence. $\delta HfH_{1.68}$ indicated lower κ_{el} than that of pure Hf. On the other hand, the κ_{lat} of both $\delta HfH_{1.68}$ and pure Hf gradually decreased with temperature and $\delta HfH_{1.68}$ indicated higher κ_{lat} than that of pure Hf. For pure Hf, the κ_{el} was dominant in the whole temperature range, while for $\delta HfH_{1.68}$, κ_{lat} was dominant at low temperature below around 450 K and at higher temperature the κ_{el} became predominant. Although, $\delta HfH_{1.68}$ and pure Hf indicated similar values of total κ, it was revealed that the components of κ were completely different.

Figure 7 shows the room temperature values of the Young's (E) and shear (G) moduli of pure Hf and δHfH_x, as a function of C_H. The E and the G were calculated from the measured longitudinal (V_L) and shear (V_S) sound velocities at room temperature. The sound velocities of the samples decreased with increasing C_H. In this figure, the literature values for pure Hf [18] were also plotted for comparison. A good agreement between the present data and the literature data [18] for pure Hf can be confirmed. Both the E and G of Hf hydride decreased with increasing C_H. In the previous studies on Ti hydride [19] and Zr hydride [20], the similar tendency has been observed. In order to understand this behavior, the electronic structure of Zr hydride has been investigated by means of the x-ray photoelectron spectroscopy and the first principle molecular orbital calculation [21-23]. It has been proposed in Refs. 21, 22, and 23 that the valence electron transfer from Zr 4d band to the Zr-H bonding state occurs in Zr hydride to induce the reduction of the Zr-Zr metallic bond with increasing C_H, leading to decrease the elastic moduli. In the Hf hydride system, the similar discussion would be applicable to account for the tendency of the decrease in the elastic moduli with increasing C_H.

Figure 8 shows the room temperature values of the Vickers hardness (H_V) of pure Hf and δHfH_x, as a function of C_H. As in the case of E, the H_V values of Hf hydride were higher than that of pure Hf and decreased with increasing C_H. The effect of hydrogenation not only on the elastic moduli but also on the hardness of Hf hydride was very similar to those of the related metal hydrides [19,20].

SUMMARY

The polycrystalline dense Hf hydride samples were prepared in which the hydrogen content C_H (H/Hf ratio) ranges from 1.62 to 1.72, and the lattice parameter (a), heat capacity (C_P), and thermal conductivity (κ), Young's modulus (E), shear modulus (G), and Vickers hardness (H_V) were examined. The XRD analysis revealed that all the prepared samples indicated single phase of cubic fluorite-type structure (delta-phase). The authors succeeded in evaluating the effect of C_H and/or temperature on the a, C_P, κ, E, G, and H_V of Hf hydride. The data obtained in the present study would be very useful in evaluating safety and reliability of advanced FRs in which Hf hydride will be used as long-life neutron control materials with no helium production.

ACKNOWLEDGEMENTS

Present study includes the result of "Development Study of Fast Reactor Core with Hydride Neutron Absorber" entrusted to Tohoku University by the Ministry of Education, Culture, Sports, Science and Technology of Japan (MEXT).

REFERENCES

1. K. Konashi, T. Iwasaki, T. Terai, M. Yamawaki, K. Kurosaki, K. Itoh, *Proceedings of the*

2006 International Congress on Advances in Nuclear Power Plants, Embedded Topical Meeting, Reno, NV, United States, June 4-8, 2006, 2213-2217 (2006).

2. D. Setoyama, J. Matsunaga, M. Ito, H. Muta, K. Kurosaki, M. Uno, S. Yamanaka, *J. Nucl. Mater.*, **344**, 298-300 (2005).

3. S. Yamanaka, K. Yamada, K. Kurosaki, M. Uno, K. Takeda, H. Anada, T. Matsuda, S. Kobayashi, *J. Nucl. Mater.*, **294**, 94-98 (2001).

4. M. Ito, K. Kurosaki, H. Muta, M. Uno, K. Konashi, S. Yamanaka, *J. Nucl. Sci. Technol.*, **46**, 814-818 (2009).

5. M. Ito, K. Kurosaki, H. Muta, M. Uno, K. Konashi, S. Yamanaka, *J. Nucl. Sci. Technol.*, **47**, 156-159 (2010).

6. S. S. Sidhu, L. R. Heaton, D. D. Zauberis, *Acta Cryst.*, **9**, 607-614 (1956).

7. S. S. Sidhu, J. C. McGuire, *J. Appl. Phys.*, **23**, 1257-1261 (1952).

8. L. Espagno, P. Azou, P. Bastien, *Compt. rend.*, **250**, 4352-4354 (1960).

9. O. M. Katz, J. A. Berger, *Trans. AIME*, **233**, 1014 (1965).

10. F. A. Lewis, A. Aladjem, *Hydrogen Metal Systems*, Scitec Publications Ltd., Balaban Publishers, Zürich, Switzerland (1996).

11. J. D. Cox, D. D. Wagman, V. A. Medvedev, *CODATA Key Values for Thermodynamics*, Hemisphere, New York, (1989).

12. B. Tsuchiya, M. Teshigawara, K. Konashi, S. Nagata, T. Shikama, *J. Alloys Compd.*, **446-447**, 439-442 (2007).

13. Y. S. Touloukian, *Thermal diffusivity*, Thermophysical Properties of Matter, 10, IFI/Plenum, New York, (1973).

14. R. A. Andrievskii, V. I. Savin, R. A. Lyutikov, *Zhurnal Neorganicheskoi Khimii*, **17**, 915 (1972).

15. P. L. Rossiter, *The electrical resistivity of metals and alloys*, Cambridge solid state science series, Cambridge University Press, Cambridge; New York, (1991).

16. M. Uno, K. Yamada, T. Maruyama, H. Muta, S. Yamanaka, *J. Alloys Compd.*, **366**, 101-106 (2004).

17. M. Ito, D. Setoyama, J. Matsunaga, H. Muta, K. Kurosaki, M. Uno, S. Yamanaka, *J. Alloys Compd.*, **420**, 25-28 (2006).

18. T. Gorecki, *Mater. Sci. Eng.*, **43**, 225-230 (1980).

19. D. Setoyama, J. Matsunaga, H. Muta, M. Uno, S. Yamanaka, *J. Alloys Compd.*, **381**, 215-220 (2004).

20. S. Yamanaka, K. Yoshioka, M. Uno, M. Katsura, H. Anada, T. Matsuda, S. Kobayashi, *J. Alloys Compd.*, **293-295**, 23-29 (1999).

21. S. Yamanaka, K. Yamada, K. Kurosaki, M. Uno, K. Takeda, H. Anada, T. Matsuda, S.

Kobayashi, *J. Alloys Compd.*, **330-332**, 99-104 (2002).

22. S. Yamanaka, K. Yamada, K. Kurosaki, M. Uno, K. Takeda, H. Anada, T. Matsuda, S. Kobayashi, *J. Alloys Compd.*, **330-332**, 313-317 (2002).

23. T. Nishizaki, S. Yamanaka, *Trans. At. Energy Soc. Japan*, **1**[4], 328-334 (2002). [in Japanese]

Mater. Res. Soc. Symp. Proc. Vol. 1215 © 2010 Materials Research Society   1215-V18-04

Oxygen Potential of $(Th_{0.7}Ce_{0.3})O_{2-x}$

Masahiko Osaka[1], Kosuke Tanaka[1], Shuhei Miwa[1],
Ken Kurosaki[2], Masayoshi Uno[3] and Shinsuke Yamanaka[2]
[1]Oarai Research and Development Center, Japan Atomic Energy Agency, 4002 Narita-cho, Oarai-machi, Higashiibaraki-gun, Ibaraki, 311-1393, Japan.
[2]Division of Sustainable Energy and Environmental Engineering, Graduate School of Engineering, Osaka University, 2-1 Yamadaoka, Suita-shi, Osaka, 565-0871, Japan.
[3]Division of Nuclear Power and Energy Safety Engineering, Graduate School of Engineering, Fukui University, 3-9-1 Bunkyo, Fukui-shi, Fukui, 915-8507, Japan.

ABSTRACT

Oxygen potentials of $(Th_{0.7}Ce_{0.3})O_{2-x}$ were experimentally determined by means of thermogravimetric analysis as a function of non-stoichiometry at 1173 and 1273 K. Oxygen potentials of $(Th_{0.7}Ce_{0.3})O_{2-x}$ at each temperature increased with increase of oxygen-to-metal (O/M) ratio ($=2-x$) and steep increases of the oxygen potentials when approaching O/M ratio $= 2$ were observed. These characteristics are typical for non-stoichiometric fluorite-type actinide dioxides. The oxygen potentials of $(Th_{0.7}Ce_{0.3})O_{2-x}$ were similar to those of CeO_{2-x} when they were plotted as a function of average Ce valence.

INTRODUCTION

ThO_2-based oxides have attracted special attention for use in various nuclear fuel systems owing to their superior properties. For example, a solid solution of $(Th,Pu)O_{2-x}$ is being investigated as a fast breeder reactor fuel in India. ThO_2 is also a promising matrix as a radioactive waste form. The present authors have been investigating a ThO_2-based oxide as a novel fuel form dedicated to minor actinide incineration in a fast reactor system [1]. On the other hand, ThO_2 has been investigated as a solid electrolyte since it was found that notable oxygen ion transportation occurred when aliovalent cations were doped to form oxygen vacancies [2]. Namely, when two trivalent rare earth atoms such as Nd are doped and replaced Th ions, one oxygen vacancy is formed to maintain electrical neutrality. Thus, oxygen non-stoichiometry occurs for the ThO_2-based oxides. This is of particular importance for ThO_2-based oxide nuclear fuels since large amounts of fission products and actinides can dissolve into the ThO_2 matrix to form oxygen non-stoichiometry [3]. Oxygen potential, which is a function of oxygen non-stoichiometry and temperature, of ThO_2-based solid solutions should be determined prior to property measurements since the oxygen non-stoichiometry greatly affects various properties that are indispensable for fuel design, such as thermal conductivity, lattice parameter and so on.

In this study, oxygen potentials of $(Th,Ce)O_{2-x}$ were experimentally determined at 1173 and 1273 K and were discussed in terms of ThO_2 effects on the oxygen potential of CeO_{2-x}. In particular, focus is placed on how the oxygen potential of solute, CeO_{2-x}, was increased or decreased by solving it into ThO_2 matrix. Such a characteristic is technologically valuable in terms of not only the extreme importance of oxygen potential itself as a fundamental property but also because CeO_{2-x} can serve as a representative of several important actinide dioxides including PuO_{2-x} and AmO_{2-x}.

EXPERIMENT

A $(Th_{0.7}Ce_{0.3})O_{2-x}$ solid solution sample was prepared by a conventional powder metallurgical route. Appropriate amounts of ThO_2 and CeO_2 powders were weighed and thoroughly mixed in acetone medium in an agate mortar using a pestle. The ThO_2 powder was from Nakarai Chemicals, Ltd. The CeO_2 powder was from Wako Pure Chemical Industries, Ltd. (99.9 % purity). The mixed powder was then compacted into a columnar pellet by a uni-axial pressing unit at 70 MPa. The compacted pellet was heated at 1673 K for 15 h under air atmosphere. The heat-treated pellet was ground, thoroughly mixed and milled again. This sample preparation procedure was repeated 3 times in order to form a uniform solid solution. Temperatures and heating times for the second and third heat-treatments were changed to 1773 K, 20 h and 1873 K, 40 h, respectively. X-ray diffraction analysis of the final heat-treated sample showed formation of a single phase solid solution with a fluorite-type structure, $(Th_{0.7}Ce_{0.3})O_{2-x}$.

About 200 mg of sample was subjected to thermogravimetric analysis (TGA) for the determination of oxygen-to-metal ratio (O/M ratio, which corresponds to 2.00-x), as a function of oxygen partial pressure, p_{O_2}, at pre-determined temperatures, specifically 1173 and 1273 K. A Rigaku TGA apparatus (model TG-8120) connected with a gas supply system was used. The $(Th_{0.7}Ce_{0.3})O_{2-x}$ sample was loaded into an alumina pan and placed in the TGA apparatus. For reference, an alpha-alumina sample was loaded into another alumina pan and also placed in the apparatus. Equilibrium p_{O_2} was adjusted in the range from 10^{-23} to 10^{-17} MPa at 1173K and from 10^{-21} to 10^{-15} MPa at 1273K by changing the ratio of H_2O to H_2 in the flowing gas. Equilibrium p_{O_2} in the outlet gas flow was measured with a stabilized zirconia oxygen sensor. The oxygen sensor was calibrated prior to the TGA using the oxidation reaction of pure chromium metal and a standard gas containing a known amount of oxygen. Microgram order weight changes in the sample were continuously monitored while changing p_{O_2} at the pre-determined temperatures step by step. Oxygen-to-metal ratios at various p_{O_2} at the pre-determined temperatures were calculated from the weight changes relative to the stoichiometry, O/M = 2.00. Oxygen partial pressures that give the stoichiometry at each temperature were defined as those at $p_{O_2}=10^{-3}$ MPa [4], which was obtained by using dilute O_2 gas.

RESULTS AND DISCUSSION

Figure 1 shows oxygen potentials of $(Th_{0.7}Ce_{0.3})O_{2-x}$ at 1173 and 1273 K as a function of O/M ratio. Oxygen potential, ΔG_{O_2}, is defined as $\Delta G_{O_2}=RT \ln p_{O_2}$, where R is the gas constant and

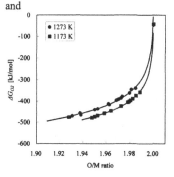

Figure 1. Oxygen potentials of $(Th_{0.7}Ce_{0.3})O_{2-x}$ at 1173 and 1273 K as a function of O/M ratio.

200

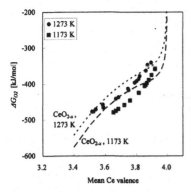

Figure 2. Oxygen potentials of $(Th_{0.7}Ce_{0.3})O_{2-x}$ at 1173 and 1273 K as a function of mean Ce valence.

T is temperature. The ΔG_{O_2} values steeply increase as O/M ratios approach stoichiometry at both temperatures. The ΔG_{O_2} values are about 30 kJ/mol higher at 1273 K than at 1173 K throughout the measured O/M ratio ranges. These features are common for non-stoichiometric actinide dioxides.

Figure 2 compares ΔG_{O_2} of $(Th_{0.7}Ce_{0.3})O_{2-x}$ with those of CeO_{2-x} [5] as a function of mean Ce valence. The chemical form of $(Th_{0.7}Ce_{0.3})O_{2-x}$ can be written as $(Th_{0.7}Ce^{4+}_{0.3-2x}Ce^{3+}_{2x})O_{2-x}$ considering possible valency of each cation in the solid. Direct comparison of ΔG_{O_2} can thus be made between pure CeO_{2-x} and ThO_2 solid solution by assuming the ΔG_{O_2} is such a function [6]. It is seen that ΔG_{O_2} values of $(Th_{0.7}Ce_{0.3})O_{2-x}$ are comparable to those of pure CeO_{2-x}. This indicates that the ΔG_{O_2} values of CeO_{2-x} are hardly affected by dissolving it into ThO_2 matrix. Namely, in terms of thermochemistry, there can be little interaction between CeO_{2-x} and ThO_2.

Nevertheless, the ΔG_{O_2} values of $(Th_{0.7}Ce_{0.3})O_{2-x}$ are somewhat lower than those of CeO_{2-x} in the measured O/M ratio ranges at both temperatures. Additionally, the shape of the ΔG_{O_2} isotherms of $(Th_{0.7}Ce_{0.3})O_{2-x}$ differ from those of CeO_{2-x}. These differences are attributed to some interactions between CeO_{2-x} and ThO_2 based on the assumption that $(Th_{0.7}Ce_{0.3})O_{2-x}$ is a solid solution of CeO_{2-x} and ThO_2. In order to see these interactions between ThO_2 and CeO_{2-x}, the activity coefficient γ of CeO_{2-x} in $(Th_{0.7}Ce_{0.3})O_{2-x}$ solid solution was calculated using the following equation [6].

$$RT \ln \gamma = \frac{1}{2y} \int_0^{yz} \Delta G_{O_2}[(Th_{0.7}Ce_{0.3})O_{2-yz}]d(yz) + \frac{1}{2} \int_z^0 \Delta G_{O_2}[CeO_{2-z}]dz \qquad (1)$$

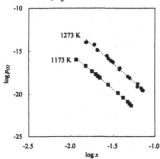

Figure 3. Log of oxygen partial pressure vs. deviation from stoichiometry for $(Th_{0.7}Ce_{0.3})O_{2-x}$.

201

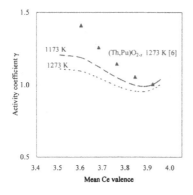

Figure 4. Activity coefficient γ of CeO_{2-x} in $(Th_{0.7}Ce_{0.3})O_{2-x}$.

This equation is based on the assumption that CeO_{2-x} and ThO_2 are considered a solute and solvent, respectively. Analytical expressions, which are shown as solid lines in Figure 1, were used for ΔG_{O_2} of $(Th_{0.7}Ce_{0.3})O_{2-x}$ in equation (1) based on slopes of $\log p_{O2}$ versus $\log x$. The slopes suggest a predominant defect structure in $(Th_{0.7}Ce_{0.3})O_{2-x}$ and the slopes remain unchanged unless the predominant defect structure is changed for such reasons as interactions between oxygen vacancies and dopant cations [7]. The analytical values of ΔG_{O_2} in the measured O/M ratio ranges can therefore be considered as reasonable since the slope values are constant as shown in Figure 3. Figure 4 shows changes of activity coefficients as a function of mean Ce valence, together with those of $(Th,Pu)O_{2-x}$ [6]. The higher the temperature is, the closer the activity coefficients becomes to 1, which means the closer it comes to the ideal solid solution. The activity coefficients become higher and depart from 1 when the mean Ce valence becomes lower. This tendency suggests that the interaction becomes larger when the deviation from stoichiometry becomes larger. Nevertheless, the activity coefficients of $(Th_{0.7}Ce_{0.3})O_{2-x}$ are smaller than those of $(Th_{0.7}Pu_{0.3})O_{2-x}$ as shown in Figure 4. Thus, $(Th_{0.7}Ce_{0.3})O_{2-x}$ can be regarded as a nearly ideal solid solution of CeO_{2-x} and ThO_2.

CONCLUDING REMARKS

Oxygen potentials of $(Th_{0.7}Ce_{0.3})O_{2-x}$ were experimentally determined by means of thermogravimetric analysis as a function of non-stoichiometry at 1173 and 1273 K. The oxygen potentials of $(Th_{0.7}Ce_{0.3})O_{2-x}$ were similar with those of CeO_{2-x} when they were plotted as a function of average Ce valence in the solid solution. This indicated that the oxygen potentials of $(Th_{0.7}Ce_{0.3})O_{2-x}$ were derived from those of CeO_{2-x}. The activity coefficient analysis indicated that $(Th_{0.7}Ce_{0.3})O_{2-x}$ could be regarded as a nearly ideal solid solution of ThO_2 and CeO_{2-x}. The present results implied an inertia feature of ThO_2 toward CeO_{2-x}.

REFERENCES

[1] M. Osaka, M. Koi, S. Takano, Y. Yamane and T. Misawa, J. Nucl. Sci. Technol. 43, 367 (2006).
[2] A. Hammou, C. Deportes, G. Robert and G. Vitter, Mat. Res. Bull. 6, 823 (1971).
[3] C. Keller, U. Berndt, H. Engerer and L. Leitner, J. Solid State Chem. 4, 453 (1972).

[4] M. Osaka, J. Alloys Compd. 475, L31 (2009).
[5] T. B. Lindemer, CALPHAD 10, 129 (1986).
[6] R. E. Woodley, J. Nucl. Mater. 96, 5 (1981).
[7] O. T. Sorensen, Thermodynamics and defect structure of nonstoichiometric oxides, in: *Nonstoichiometric Oxides*, edited by O. T. Sorensen (Academic Press Inc., New York, 1981), pp. 1-59.

Thermal Aging of Primary Circuit Piping Materials in PWR Nuclear Power Plant

Xitao Wang [1)], Shilei Li [1)], Shuxiao Li [1)], Yanli Wang [1)], Fei Xue [2)], Guogang Shu [2)]

[1] State Key Laboratory for Advanced Metals and Materials, University of Science and Technology Beijing, 30 Xueyuan Rd. , Beijing 100083, China
[2] China Guangdong Nuclear Power Group, Shenzhen 518028, China

ABSTRACT

The reserved cast austenitic stainless steels (CASS) for primary circuit piping in Daya Bay Nuclear Power Plant were studied. The changes of microstructure, mechanical properties and fracture behavior were investigated using SEM, EPMA, TEM and nanoindentation after accelerated aging at 400°C for up to 10000 h. Microhardness of ferrite increased rapidly in the early stage and then increased slowly later. The impact energy of materials declined with the aging time and reduced to a very low level after aging for 10000 hours. Fracture morphology displayed a mixture of cleavage in ferrite along with dimple and tearing in austenite. Two kinds of precipitations were observed in ferrite by TEM after long periods of aging. The fine Cr-enriched α' phases precipitated homogeneously in ferrite, and a few larger G phases were observed as well. The precipitation of α' phases was considered to be the primary mechanism of thermal aging embrittlement in CASS.

INTRODUCTION

CASS materials are widely used in major components of PWR nuclear power plants due to their excellent strength, corrosion resistance and good weldability. The microstructure of CASS contains about 10-20% island ferrite in an austenite matrix. When they are used in piping system of primary cooling circuits, the service temperature is at the intermediate range of 280°C ~320°C, which is under its ductile-brittle transition temperature. However, when they are used in this temperature range for extended periods of time, they can suffer a loss of toughness [1]. Because no changes were observed in the austenite phases after long term aging, the embrittlement was considered to be associated with the changes in the ferrite phases [2]. Microstructure changes in the ferrite phases, for this type of aged cast stainless steels, were characterize by Chung and Chopra[3,4], Yamada[5], Auger[6], et al. Microstructural studies of these materials reported that phase decomposition to be the principal mechanism, wherein the ferrite decomposes into Fe-rich α and Cr-rich α'. G-phase precipitation in the ferrite was reported as secondary reactions [7-10].

Because realistic aging of component for end-of-life conditions at service temperature could not be produced, it was customary to simulate the metallurgical structure of a reactor component by accelerated aging at 400°C. Relative studies[11,12] indicated that the mechanisms of aging embrittlement were identical for the accelerated aging and reactor operating conditions.

EXPERIMENTAL DETAILS

The material studied in this work is centrifugally CASS, containing 12% ferrite phase, which was cut from the primary coolant water pipe in Daya Bay Nuclear Power Plant. The chemical composition of this material is listed in Table 1. The material was cut into small pieces and aged at 400°C for different aging time using electric furnace in air.

Table 1. Chemical composition of the material (wt %)

C	Si	Mn	P	S	Cr	Ni	Mo	Cu	Co	N
0.027	1.27	1.13	0.023	0.014	20.19	8.92	0.21	0.094	0.044	0.031

The corresponding microstructure, hardness and toughness were determined. Charpy impact tests were carried out at room temperature. The hardness tests were made by using a Vickers hardness tester (with a load of 25 g) and a nanoindenter. Fractography of the impact specimens was performed in a ZEISS SUPRA55 SEM operated at 15kV. TEM specimens were sliced from the aged specimens, thinned to 0.06mm by abrasion on SiC papers and twin-jet electropolished using 95% ethanol 5% perchloric acid electrolyte at 20-40V. Carbon extraction replicas were prepared from the material aged at 400°C for 10000h. The specimens were etched electrolytically in 10% HCl solution in ethyl alcohol at 10 V prior to shadowing by carbon. The microstructure of aged specimens was investigated using a JOEL JEM 2000 EX II equipped with an EDS.

RESULTS AND DISCUSSION

Microstructure analysis

The microstructure of as-received and aged material is shown in Figure 1. No change of microstructure in micro-scale has been observed after aging for 10000 h.

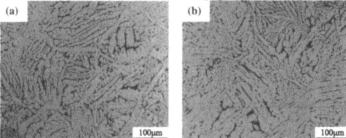

Figure 1. Backscattered electron image of CASS (a) As-received; (b) aged at 400°C for 10000h

Alloying elements analysis

The alloying elements contents in the ferrite and austenite analyzed by EPMA are shown in Table 2. Chromium, silicon and molybdenum are enriched in ferrite phase, and nickel, manganese, copper and cobalt are enriched in the austenite phase. The elements contents in the same phase of the as-received and aged for 10000 h are basically identical.

Table 2. Alloying elements contents analyzed by EPMA (wt %)

	Aging time/h	Fe	Cr	Ni	Si	Mn	Mo	Cu	Co	Nb
Ferrite	0	65.459	27.626	5.624	1.141	0.935	0.173	0.046	0.073	0.036
	10000	65.144	27.508	5.618	1.270	0.895	0.171	0.034	0.080	0.015
Austenite	0	67.515	20.674	10.257	0.998	1.077	0.114	0.072	0.114	0.031
	10000	67.973	20.948	9.866	1.101	1.064	0.131	0.073	0.103	0.017

Hardness analysis

The aging behaviors of ferrite and austenite were evaluated by the 25g Vickers microhardness test. The large ferrite islands (>40 μm) were measured to eliminate the influence of the softer underlying austenite on thin ferrite islands. The microhardness values obtained as the mean of five measurements are shown in Figure 2(a). The austenite hardness does not change with aging time. In contrast, the ferrite hardness markedly increases in the early stage and then increases slowly later. Figure 2(b) shows the nanohardness of ferrite obtained by nanoindentation change with aging time. The nanohardness continuously increases in the aging process, showing different change trend with microhardness.

Precipitation strengthening in the ferrite phases causes the increase of ferrite hardness. In the early stage of aging, the precipitation of a large amount of particles in ferrite lead to the strong precipitation hardening effect, which result in the significantly increase of microhardness and nanohardness. But in the later stage, the growth and coarsening of the precipitates have not so strong hardening effect. The indentation size effect might be the reason of the difference in the variation between the two kinds of hardness.

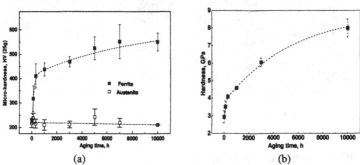

(a)   (b)

Figure 2. Microhardness (a) and nanohardness (b) of ferrite change versus aging time.

Charpy impact test

Figure 3(a) shows the change in the Charpy impact energy versus aging time at 400°C. The Charpy impact energy decrease rapidly with aging time, the loss of impact energy does not saturate even after aging at 400°C for 10000h. Ferrite hardness markedly increases, whereas austenite hardness does not change with aging. There is a possible correlation between the loss of Charpy impact energy and ferrite hardness. Figure 3(b) shows the relationship between the Charpy impact energy and ferrite hardness. The Charpy impact energy decreases with increasing ferrite microhardness in a linear relation.

The Charpy impact test was carried out at room temperature. Figure 4 shows the fracture morphology, which displayed a mixture of cleavage in ferrite along with dimple and tearing in austenite.

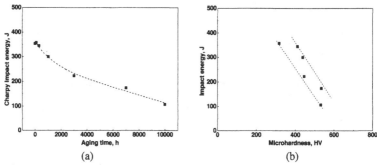

(a)   (b)

Figure 3. (a) Charpy impact energy changes of materials aged at 400°C with aging time.
(b) Relationship between the Charpy impact energy and ferrite microhardness.

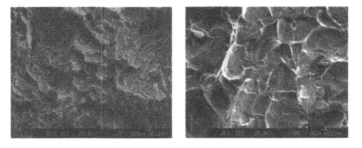

Figure 4. Fracture morphology of the material aged for 10000 h.

Transmission electron microscopy observation

The microstructure of ferrite region in aged materials at 400°C for 10000 h is shown in Figure 5(a). As shown in the TEM image, there are fine precipitates through the whole ferrite, 3~5 nm in diameter. Selected area diffraction (SAD) analysis of a random area in ferrite phase showed only one set of b.c.c spots. This is to be expected for b.c.c precipitates whose lattice parameter is almost identical to that of the b.c.c matrix. The TEM image of carbon extraction

208

replica of the material aged at 400°C for 10000 h is shown in Figure 5(b). Two phases are recognized, one is fine particles shown as A, and the second with a larger size of 50 nm , shown as B.

The results of the EDS analysis of these two kind precipitates on the carbon extraction replica are shown in Figure 6. The composition of finer precipitates shown as A is 7.77% Si, 65.10% Cr, 0.35% Mn, 14.98% Fe, 7.56% Ni, and 4.24% Mo. From the high Cr content and the low Fe content, combining the results of SAD analysis, we could conclude the finer precipitates are Cr-rich α' phase. Chemical composition of coarse particles shown as B is 29.59% Si, 35.06% Cr, 18.83% Fe, and 16.52% Mo. The content of Si and Mo indicate that the phase is most probably G-phase.

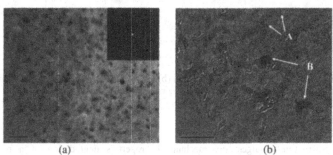

(a)   (b)

Figure 5. (a) TEM image and SAD pattern (b) carbon extraction replica image of the material aged for 10000 h.

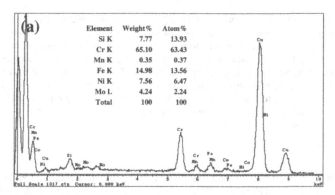

Element	Weight%	Atom%
Si K	7.77	13.93
Cr K	65.10	63.43
Mn K	0.35	0.37
Fe K	14.98	13.56
Ni K	7.56	6.47
Mo L	4.24	2.24
Total	100	100

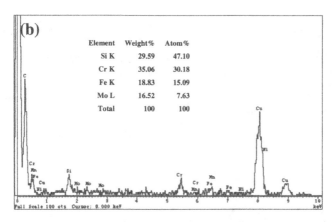

Element	Weight%	Atom%
Si K	29.59	47.10
Cr K	35.06	30.18
Fe K	18.83	15.09
Mo L	16.52	7.63
Total	100	100

Figure 6. EDS analysis of (a) fine particles and (b) coarse particles in carbon extraction replica.

CONCLUSIONS

Microstructure changes and mechanical properties degradation of a reserved cast austenitic stainless steel for primary circuit piping in Daya Bay Nuclear Power Plant from Daya Bay Nuclear Power Plant were investigated. Thermal aging is accelerated at 400°C for up to 10000 hours.

The aged steel exhibits a rapid ferrite hardening within the first 300 hours, and then a slower hardening stage up to 10000h at 400°C. The room temperature Charpy impact energy of materials decreases with aging time and reduced to a very low level after 10000 hours. However, the loss of impact energy has not saturated. Charpy impact energy of aged material decreases linearly with increasing ferrite hardness. Fracture morphology of the long time aged materials displayed a mixture of cleavage in ferrite along with dimple and tearing in austenite. TEM analysis of aged material in thin foils and carbon extraction replica revealed the presence of large amount 5nm fine Cr-enriched α' phase distributed homogeneously in ferrite phase. A few larger G phases in about 50 nm size were also observed.

ACKNOWLEDGMENTS

The financial support by the National Basic Research Program of China (973 Program, No.2006CB605005), the National High-Tech Research and Development Program of China (863 Program, No.2008AA031702) and China Guangdong Nuclear Power Group are greatly acknowledged.

REFERENCES

[1] H M Chung,T R Leax. *Materials Science and Technology*, 6,249(1990).

210

[2] O K Chorpa. Argonne: Argonne National Laboratory, CONF-950908-10(1995).

[3] H M Chung. *International Journal of Pressed Vessels and Piping*, 50,179(1992).

[4] H M Chung Argonne: Argonne National Laboratory, ANL/CP-70872(1991).

[5] T Yamada, S Okano, H Kuwano. *Journal of Nuclear Materials*, 350,47(2006).

[6] F Danoix, P Auger. *Materials Characterization*, 44,177(2000).

[7] M Vrinat, Cozar R, Meyzaud Y.*Scripta Metallurgica*, 20,1101(1986).

[8] M D Mathew, L M Lietzan, K L Murty, et al. *Materials Science and Engineering* A, 269, 186 (1999).

[9] F Danoix, P Auger, D Blavette. *Microscopy and Microanalysis*,10,349(2004).

[10] S Kawaguchi, N Sakamoto, G Takano, et al. *Nuclear Engineering and Design*, 174, 273 (1997).

[11] O K Chorpa. Argonne: Argonne National Laboratory, CONF-8310143-65(1983).

[12] O K Chorpa. Argonne: Argonne National Laboratory, CONF-8410142-38(1984).

AUTHOR INDEX

SUBJECT INDEX